"Why do engineers prefer GLP study guides?"

Reader remarks from the GLP in-basket....

"The multiple-choice and essay sample problems in your PE review were exactly like what I found when I took the actual exam!"

—John Jurgensen, Louisiana

"In your review books you cut to the chase. Yours is the only review that doesn't scare the pants off people. It's exactly what I want for teaching my review course."

—John Thorington

"I passed the first time I took the test. Potter's book is a great, great review. It's right to the point and doesn't include extra stuff you don't need. All I did was work through the book. My friends took review courses and didn't use your book—they failed! Do you have a PE review? I don't want to take the exam until you do!"

—Dana, calling to order our PE Civil review

"I used your FE book and thought it was far superior to any other. I took the exam and did great. Now I need your PE Review!"

—Tom Roach, Indiana Midwest Steel

"Your *Principles & Practice of Civil Engineering* review was well worth the money spent! It was extremely easy to use during the exam and greatly increased my comfort level. The chapters on soils, hydrology, and open channel flow are exceptional."

—Mark MacIntire, Virginia

"When you're working 50 hours a week, you don't have a lot of extra time to prepare. Potter books are condensed and focus on only what is on the exam. It's a good strategy to concentrate on what is on the exam—just what you need to pass. It doesn't overwhelm you."

—James Alt, graduate of University of Wisconsin

"I used your FE review the second time I took the exam and finally passed it! So, obviously I turned to your PE review and was equally pleased. It is informative, to-the-point, and covers all the material needed for the exam. Unlike other reviews, yours are easy to follow and I don't have to dig through anything that is not on the exam."

—Gena Swift, Illinois

"Several of the engineers I sold a copy of the FE book to passed the exam after trying and failing repeatedly. They all said "What is it with this book?" They didn't even attend my course and they passed with the book alone."

—Shahin Mansour, S&R Engineering Foundation

International Standard Book Number 1-881018-70-9

All comments and inquiries should be addressed to:
Great Lakes Press—Customer Service
PO Box 550
Grover, MO 63040-0550
Phone: (314) 273-6016
Fax: (314) 273-6086
Website: www.glpbooks.com
Email: custserv@glpbooks.com

Library of Congress
Cataloging-in-Publication Data

Printed in the USA by Braun-Brumfield, Inc., Ann Arbor, Michigan.

10 9 8 7 6 5 4 3 2 1

PRINCIPLES & PRACTICE OF MECHANICAL ENGINEERING

Solutions Manual

...to the Practice Problems of each chapter...

Editor: MERLE C. POTTER, PhD, PE
Professor, Michigan State University

Authors:		
	Frank Hatfield, PhD, PE	Engineering Economics
	George E. Mase, PhD	Mechanics
	Merle C. Potter, PhD, PE	Thermodynamics
	Charles W. Radcliffe, PhD	Dynamics
	Clark J. Radcliffe, PhD	Design
	Bassem H. Ramadan, PhD	HVAC
	Craig W. Somerton, PhD	Power Plants
	David C. Wiggert, PhD, PE	Fluid Flow
	Keith Woodbury, PhD	Heat Transfer

The authors are professors at Michigan State University, with the exception of K. Woodbury, of The University of Alabama and C. W. Radcliffe, from The University of California-Berkeley

published by:

GREAT LAKES PRESS

P.O. Box 483
Okemos, MI 48805-0483
Customer Service (314) 273-6016
Fax (314) 273-6086
www.glpbooks.com   custserv@glpbooks.com

Table of Contents

Solutions to Practice Problems

Preface 1
Mathematics 3
Statics 9
Dynamics 15
Mechanics of Materials 21
Thermodynamics 33
Fluid Mechanics 39
Kinematics and Dynamics of Machinery 47
Stress Analysis 59
Vibration 65
Machine Design 71
Control Systems 77
Heat Transfer 87
Heating, Ventilation, Air Conditioning and Refrigeration 95
Power Plants 101
Fluid Flow 109
Engineering Economics 117

Preface

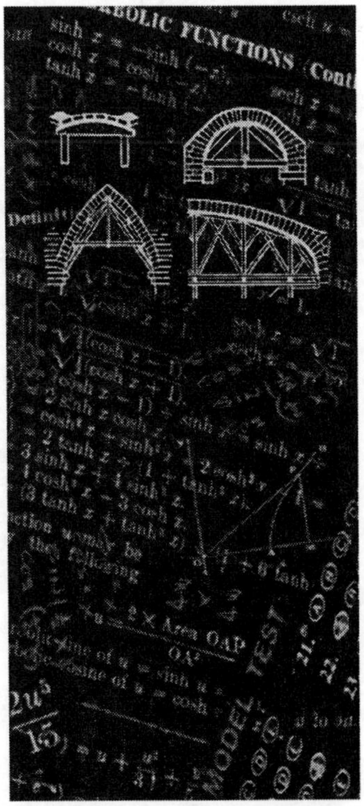

This solutions manual provides the solutions to the practice problems at the end of each chapter in the *Principles & Practice of Mechanical Engineering Review,* second edition. Most of the solutions are intentionally not as detailed as the examples in the book; they are intended to provide only the primary steps in each solution showing the appropriate numbers substituted into the required equation resulting in the answer. Units are sometimes displayed if they are not obvious. Occasionally, subtle points may be discussed. In general, some knowledge of a problem may be needed to fully understand all the steps presented; this knowledge is assumed to come from the presentation and the examples worked out in detail in the review material.

The degree of difficulty and length of solution for each problem varies considerably. Some are relatively easy and others are quite difficult. This parallels the actual exam. Typically, the easier problems come first, especially in the multiple-choice questions.

The examples in the book and the solutions in this manual have been carefully solved with the hope that errors have not been introduced. Even though extreme care is taken and problems are worked repeatedly, errors do creep in. We would appreciate knowing about any errors that you may find. They can be eliminated in future printings. Suggestions for changes and improvements are also welcomed. Write to: GLP, POB 550, Grover, MO 63040, send email to: custserv@glpbooks.com, fax: 314-273-6086.

Why a separate solutions manual?

It is necessary to place the solutions in a separate book because some states do not allow solutions to exam-like problems into the exam room, where our review book is meant to be used. One can either check with one's state board to decide if this manual is allowed in the exam room, or it can be left outside the exam room if it is not allowed. Such decisions depend on individual interpretations of proctors and often vary.

Study strategy

If one is not familiar with the subjects of a sufficient number of chapters, it may be advisable to enroll in a review course where the material is presented by experts. Such a review course may be needed for a "pass" on the exam.

Since one selects four problems to work from a set of 10 problems, it is not necessary to become proficient in all subjects included in the exam. We recommend that at least four major areas are studied, assuming that the four required problems can be selected from these areas.

Dr. Merle C. Potter
Okemos, Michigan

Mathematics

by Merle C. Potter

Solutions to Practice Problems

1.1 **c)** $100 = 10e^{2t}$. $\therefore \ln e^{2t} = \ln 10$. $\therefore 2t = 2.303$ $\therefore t = 1.151$

1.2 **b)** $\ln x = 3.2$. $\therefore e^{3.2} = x$. $\therefore x = 24.53$

1.3 **c)** $\ln_5 x = -1.8$. $\therefore 5^{-1.8} = x$. $\therefore x = 0.0552$

1.4 **a)** $x = \dfrac{-(-2) \pm \sqrt{2^2 - 4(3)(-2)}}{3 \cdot 2} = 1.215$

1.5 **e)** $(4+x)^{1/2} = 4^{1/2} + \dfrac{1}{2}4^{-1/2}x + \dfrac{\frac{1}{2}\left(1-\frac{1}{2}\right)}{2}4^{-3/2}x^2 + \cdots \quad = 2 + x/4 - x^2/64 + \cdots$

1.6 **a)** $\dfrac{2}{x\left(x^2-3x+2\right)} = \dfrac{A_1}{x} + \dfrac{A_2}{x-2} + \dfrac{A_3}{x-1} = \dfrac{A_1\left(x^2-3x+2\right) + A_2\left(x^2-x\right) + A_3\left(x^2-2x\right)}{x\left(x^2-3x+2\right)}$

$$\therefore \quad \left.\begin{aligned} A_1 + A_2 + A_3 &= 0 \\ -3A_1 - A_2 - 2A_3 &= 0 \\ 2A_1 &= 2 \end{aligned}\right\} \qquad \left.\begin{aligned} A_1 &= 1 \\ A_2 + A_3 &= -1 \\ -A_2 - 2A_3 &= 3 \end{aligned}\right\} \qquad \begin{aligned} A_3 &= -2 \\ A_2 &= 1 \end{aligned}$$

1.7 **b)** $\frac{4}{x^2\left(x^2-4x+4\right)}=\frac{A_1}{x}+\frac{A_2}{x^2}+\frac{A_3}{x-2}+\frac{A_4}{(x-2)^2}$

$$=\frac{A_1\left(x^3-4x^2+4x\right)+A_2\left(x^2-4x+4\right)+A_3\left(x^3-2x^2\right)+A_4x^2}{x^2(x-2)^2}$$

$$\therefore \left.\begin{aligned} A_1+A_3&=0\\ -4A_1+A_2-2A_3+A_4&=0\\ 4A_1-4A_2&=0\\ 4A_2&=4 \end{aligned}\right\} \qquad \begin{aligned} A_2&=1\\ A_1&=1\\ A_3&=-1\\ A_4&=1 \end{aligned}$$

1.8 **d)** at $t=0$, population $=A$. $\therefore 2A=Ae^{0.4t}$. $\ln 2=0.4t$. $\therefore t=1.733$

1.9 **a)** $\sin\theta=0.7$ $\therefore\theta=44.43°$. $\tan 44.43°=0.980$

1.10 **e)** $\tan\theta=5/7$. $\therefore\theta=35.54°$

1.11 **c)** $\tan\theta\sec\theta\left(1-\sin^2\theta\right)/\cos\theta=\tan\theta\frac{1}{\cos\theta}\cos^2\theta\frac{1}{\cos\theta}=\tan\theta$

1.12 **d)** $3^2=4^2+2^2-2\cdot2\cdot4\cos\theta$. $\therefore\cos\theta=0.6875$. $\theta=46.6°$. $\therefore$ rad $=0.813$

1.13 **a)** $L^2=850^2+732^2-2\cdot850\cdot732\cos154°$. $\therefore L=1542$ m

1.14 **e)** $\cos 2\theta=\cos^2\theta-\sin^2\theta=1-\sin^2\theta-\sin^2\theta=1-2\sin^2\theta$. $\therefore 2\sin^2\theta=1-\cos 2\theta$

1.15 **b)** $\theta=\pi(n-2)/n=\pi(8-2)/8$ radians. $\frac{6\pi}{8}\times\frac{180}{\pi}=135°$

1.16 **a)** $\text{Area}=\text{Area}_{top}+\text{Area}_{sides}=\pi R^2+\pi DL$

$$=\pi\times7.5^2+\pi\times15\times10=648\text{ m}^2.\quad 648\div10\approx65$$

1.17 **a)** $y=mx+b$. $y=-2x+b$. $0=-2(2)+b$. $\therefore b=4$. $\therefore y=-2x+4$

1.18 **b)** $y=mx+b$. $0=4m+b$. $-6=b$. $\therefore m=3/2$. $\therefore y=3x/2-6$ or $2y=3x-12$

1.19 **d)** $3x-4y-3=0$. $A=3, B=-4$. $d=\frac{|3\times6-4\times8-3|}{\sqrt{3^2+(-4)^2}}=3.4$

1.20 **c)** $B^2-AC=2^2-1\times4=0$. $\therefore$ parabola

1.21 **b)** $xy=\pm k^2=4$

1.22 **c)** $\frac{x^2}{25^2}+\frac{y^2}{50^2}=1$. $\therefore 4x^2+y^2=2500$

1.23 **b)** $x = r\cos\theta = 5\times 0.866 = 4.33.\quad y = r\sin\theta = 5\times 0.5 = 2.5$

Spherical coordinates:

$$r = \sqrt{4.33^2 + 2.5^2 + 12^2} = 13.\quad \phi = \cos^{-1}\frac{z}{r} = \cos^{-1}\frac{12}{13} = 22.6°.$$

$$\theta = \tan^{-1}\frac{y}{x} = \tan^{-1}\frac{2.5}{4.33}.\quad \therefore \theta = 30°.$$

1.24 **c)** **a)** is rectangular coordinates. b) is spherical coordinates.

1.25 **a)** $\dfrac{3-i}{1+i} = \dfrac{3-i}{1+i}\dfrac{1-i}{1-i} = \dfrac{3-1-4i}{1-(-1)} = \dfrac{1}{2}(2-4i) = 1-2i$

1.26 **d)** $1+i = re^{i\theta}.\quad r = \sqrt{1^2+1^2} = \sqrt{2}.\quad \theta = \tan^{-1}\frac{1}{1} = \pi/4 \text{ rad}.\quad \therefore 1+i = \sqrt{2}e^{i\pi/4}.$

$\therefore (1+i)^6 = \left(\sqrt{2}\right)^6 e^{i\pi 3/2} = 8\left(\cos\frac{3\pi}{2} + i\sin\frac{3\pi}{2}\right) = -8i$. Note: angles are in radians.

1.27 **b)** $1+i = \sqrt{2}e^{i\pi/4}.\quad \therefore (1+i)^{1/5} = 1.414^{1/5}e^{i\pi/20} = 1.072\left(\cos\frac{\pi}{20} + i\sin\frac{\pi}{20}\right)$

$$= 1.06 + 0.168i$$

1.28 **c)** $(3+2i)(\cos 2t + i\sin 2t) + (3-2i)(\cos 2t - i\sin 2t) = 6\cos 2t + 4i(i\sin 2t)$

$$= 6\cos 2t - 4\sin 2t$$

1.29 **e)** $6(\cos 2.3 + i\sin 2.3) - 5(\cos 0.2 + i\sin 0.2) = 6(-0.666 + 0.746i) - 5(0.98 + 0.199i)$

$$= -8.90 + 3.48i$$

$$= 9.56\angle 158.6°$$

1.30 **e)** $\begin{vmatrix} 3 & 2 & 1 \\ 0 & -1 & -1 \\ 2 & 0 & 2 \end{vmatrix} = -6 - 4 + 2 = -8$

1.31 **a)** $\begin{vmatrix} 1 & 0 & 1 & 1 \\ 2 & -1 & 0 & 1 \\ 0 & 0 & 2 & 0 \\ 3 & 2 & 1 & 1 \end{vmatrix} = 2\begin{vmatrix} 1 & 0 & 1 \\ 2 & -1 & 1 \\ 3 & 2 & 1 \end{vmatrix} = 2(-1+4+3-2) = 8$

Note: Expand using the third row.

1.32 **b)** $(-1)^3\begin{vmatrix} 2 & 1 \\ 0 & 2 \end{vmatrix} = -4$

1.33 **e)** $(-1)^7\begin{vmatrix} 1 & 0 & 1 \\ 2 & -1 & 0 \\ 3 & 2 & 1 \end{vmatrix} = -(-1+4+3) = -6$

1.34 **d)** $\left[a_{ij}\right]^+ = \begin{bmatrix} A_{11} & A_{21} \\ A_{12} & A_{22} \end{bmatrix} = \begin{bmatrix} 2 & 4 \\ 0 & 1 \end{bmatrix}$

1.35 **a)** $\left[a_{ij}\right]^{-1} = \dfrac{\left[a_{ij}\right]^+}{\left|a_{ij}\right|} = \dfrac{\begin{bmatrix} 1 & -3 \\ -1 & 2 \end{bmatrix}}{-1} = \begin{bmatrix} -1 & 3 \\ 1 & -2 \end{bmatrix}$

1.36 **b)** $\begin{bmatrix} 2 & -1 \\ 3 & 2 \end{bmatrix}\begin{bmatrix} 2 \\ 1 \end{bmatrix} = \begin{bmatrix} 4-1 \\ 6+2 \end{bmatrix} = \begin{bmatrix} 3 \\ 8 \end{bmatrix}$

1.37 **a)** $\begin{bmatrix} 1 & 2 \\ 2 & 1 \end{bmatrix}\begin{bmatrix} -1 & 0 \\ 1 & 2 \end{bmatrix} = \begin{bmatrix} 1 & 4 \\ -1 & 2 \end{bmatrix}$

1.38 **d)** $\left[a_{ij}\right] = \begin{bmatrix} 3 & 2 & 0 \\ 1 & -1 & 1 \\ 4 & 0 & 2 \end{bmatrix}. \quad \left[a_{ij}\right]^+ = \begin{bmatrix} -2 & -4 & 2 \\ 2 & 6 & -3 \\ 4 & 8 & -5 \end{bmatrix}. \quad \left|a_{ij}\right| = -2.$

$\therefore \left[a_{ij}\right]^{-1} = \dfrac{\left[a_{ij}\right]^+}{\left|a_{ij}\right|} = \begin{bmatrix} 1 & 2 & -1 \\ -1 & -3 & 3/2 \\ -2 & -4 & 5/2 \end{bmatrix}. \quad \left[x_j\right] = \left[a_{ij}\right]^{-1}\left[r_i\right] = \left[a_{ij}\right]^{-1}\begin{bmatrix} -2 \\ 0 \\ 4 \end{bmatrix} = \begin{bmatrix} -6 \\ 8 \\ 14 \end{bmatrix}$

1.39 **a)** $\begin{vmatrix} 1-\lambda & 2 \\ 3 & 2-\lambda \end{vmatrix} = \lambda^2 - 3\lambda + 2 - 6 = \lambda^2 - 3\lambda - 4 = 0.$

$(\lambda-4)(\lambda+1) = 0. \quad \therefore \lambda = 4, -1$

1.40 **b)** Use $\lambda = 4$. $\begin{bmatrix} -3 & 2 \\ 3 & -2 \end{bmatrix}\begin{bmatrix} x_1 \\ x_2 \end{bmatrix} = 0$.

$-3x_1 + 2x_2 = 0.$
$3x_1 - 2x_2 = 0.$ $\quad \therefore \mathbf{x} = \begin{bmatrix} 2 \\ 3 \end{bmatrix}$

Any multiple is also an eigenvector, e.g., $\mathbf{x} = \begin{bmatrix} 1 \\ 3/2 \end{bmatrix}$ or $\mathbf{x} = \begin{bmatrix} 2/\sqrt{13} \\ 3/\sqrt{13} \end{bmatrix}$

1.41 **a)** $\dfrac{dy}{dx} = 6x^2 - 3 = 6(1)^2 - 3 = 3$

1.42 **c)** $\dfrac{dy}{dx} = \dfrac{1}{x} + e^x \cos x + e^x \sin x = 1 + e\cos 1 + e \sin 1 = 4.76.$ $\quad (\cos 1 = \cos 57.3°)$

1.43 **d)** $\dfrac{dy}{dx} = 3x^2 - 3 = 0. \quad \therefore x^2 = 1. \quad \therefore x = \pm 1. \quad \dfrac{d^2y}{dx^2} = 6x. \quad \therefore x = -1$ is a maximum.

1.44 **c)** $y' = 3x^2 - 3. \quad y'' = 6x. \quad \therefore x = 0$ is inflection.

1.45 **a)** $\lim\limits_{x\to\infty} \dfrac{2x^2 - x}{x^2 + x} = \lim\limits_{x\to\infty} \dfrac{4x-1}{2x+1} = \lim\limits_{x\to\infty} \dfrac{4}{2} = 2$

1.46 **c)** $\eta(x+h) = \eta + h\eta' + \dfrac{h^2}{2}\eta''$

1.47 **b)** $e^x \sin x = \left(1 + x + \dfrac{x^2}{2}\right)\left(x - \dfrac{x^3}{6}\right) = x + x^2 + \dfrac{x^3}{2} - \dfrac{x^3}{6} = x + x^2 + x^3/3$

1.48 **d)** $\text{Area} = \int_4^9 x\,dy = \int_4^9 y^{1/2} dy = \frac{2}{3}(27-8) = 12\frac{2}{3}$

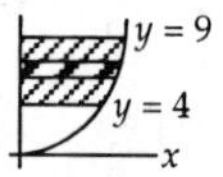

1.49 **e)** $\text{Area} = \int_0^4 (x_2 - x_1)dy = \int_0^4 \left(2y^{1/2} - \frac{y^2}{4}\right)dy$ 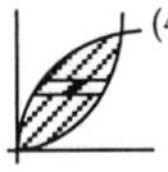

$= 2 \times \frac{2}{3} \times 8 - \frac{1}{12} \times 64 = 16/3$

1.50 **c)** $V = \int_0^2 2\pi y\, x dy = 2\pi \int_0^2 y^3 dy = 2\pi \times \frac{2^4}{4} = 8\pi$

1.51 **a)** $\int_0^2 \left(e^x + \sin x\right)dx = e^x - \cos x\Big|_0^2 = e^2 - 1 - \cos 2 + 1 = 7.81$

1.52 **a)** $\int_0^1 e^x \sin x\, dx = e^x \sin x\Big|_0^1 - \int_0^1 e^x \cos x\, dx$

$u = \sin x \quad dv = e^x dx$
$du = \cos x dx \quad v = e^x$
1st integral

$u = \cos x \quad dv = e^x dx$
$v = e^x \quad du = -\sin x dx$
2nd integral

$\therefore \int_0^1 e^x \sin x\, dx = e \sin 1 - \left[e^x \cos x\Big|_0^1 + \int_0^1 e^x \sin x\, dx\right]$

$\therefore 2\int_0^1 e^x \sin x\, dx = e \sin 1 - e\cos 1 + 1 \times 1 = 1.819. \quad \therefore \int_0^1 e^x \sin x\, dx = 0.909$

1.53 **e)** $\int x \cos x\, dx = x \sin x - \int \sin x\, dx = x \sin x + \cos x + C$

$u = x \quad dv = \cos x dx$
$du = dx \quad v = \sin x$
1st integral

1.54 **b)** linear and nonhomogeneous. The term (+2) makes it nonhomogeneous. $x^2 y'$ is linear.

1.55 **a)** $\frac{dy}{dx} = -2xy. \quad \frac{dy}{y} = -2xdx. \quad \therefore \ln y = -x^2 + C. \quad \ln 2 = 0 + C. \quad \therefore C = \ln 2.$

$y(2) = \exp\left(-2^2 + \ln 2\right) = 0.0366$

1.56 **b)** $\frac{dy}{dx} = -2x. \quad dy = -2xdx. \quad \therefore y = -x^2 + C. \quad 1 = 0 + C. \quad \therefore C = 1.$

$y(10) = -10^2 + 1 = -99$

1.57 **a)** $2m^2 + m + 50 = 0. \quad \therefore m = \frac{-1 \pm \sqrt{1 - 400}}{4} = -\frac{1}{4} \pm 4.99i.$

$\therefore y(t) = e^{-t/4}(A \cos 4.99t + B \sin 4.99t).$

$\therefore \omega = 4.99 \text{ rad/s}. \quad \therefore f = \frac{\omega}{2\pi} = \frac{4.99}{2\pi} = 0.794 \text{ hertz}$

1.58 **a)** $m^2 + 16 = 0. \quad \therefore m = \pm 4i. \quad \therefore y(t) = C_1 \cos 4t + C_2 \sin 4t$

1.59 **e)** $m^2+8m+16=0.\quad (m+4)^2=0.\quad m=-4,-4.\quad \therefore y(t)=C_1e^{-4t}+C_2te^{-4t}$

1.60 **d)** $m^2-5m+6=0.\quad m=\dfrac{5\pm\sqrt{25-24}}{2}=3,\,2.\quad \therefore y_h=C_1e^{3t}+C_2e^{2t}.$

Assume $y_p=Ae^t$. Then $Ae^t-5Ae^t+6Ae^t=4e^t.\quad \therefore A=2.$

1.61 **a)** homogeneous: $m^2+16=0.\quad \therefore m=\pm 4i.\quad \therefore y_h(t)=C_1\sin 4t+C_2\cos 4t$

particular: $y_p=At\cos 4t.$ (This is resonance.) $\dot{y}_p=A\cos 4t-4At\sin 4t$

$\therefore -4A\sin 4t-4A\sin 4t-16At\cos 4t+16At\cos 4t=8\sin 4t$

$\therefore -8A=8.\quad A=-1.\quad \therefore y=y_h+y_p=C_1\sin 4t+C_2\cos 4t-t\cos 4t$

1.62 **d)** $s^2Y-2s+5sY-10+6Y=0.\quad Y=\dfrac{2s+10}{s^2+5s+6}=\dfrac{-4}{s+3}+\dfrac{6}{s+2}.$

$\therefore y(t)=-4e^{-3t}+6e^{-2t}$

1.63 **e)** $s^2Y-12+Y=\dfrac{120}{(s+1)^2}.\quad Y=\dfrac{12}{s^2+1}+\dfrac{120}{(s+1)^2(s^2+1)}$

$$=\frac{12}{s^2+1}+\frac{60}{s+1}+\frac{60}{(s+1)^2}-\frac{60s}{s^2+1}$$

$\therefore y(t)=12\sin t-60\cos t+60\left(e^{-t}+te^{-t}\right)$

1.64 **c)** $\mathbf{A}\cdot\mathbf{B}=3\cdot 10+(-6)\cdot 4+2(-6)=-6$

1.65 **b)** $\mathbf{A}\times\mathbf{B}=(2\mathbf{i}-5\mathbf{k})\times\mathbf{j}=2\mathbf{i}\times\mathbf{j}-5\mathbf{k}\times\mathbf{j}=2\mathbf{k}-5(-\mathbf{i})=5\mathbf{i}+2\mathbf{k}$

1.66 **e)** $\mathbf{i}_\mathbf{B}=(6\mathbf{i}+3\mathbf{j}-2\mathbf{k})\big/\sqrt{6^2+3^2+2^2}=\frac{1}{7}(6\mathbf{i}+3\mathbf{j}-2\mathbf{k})$

$\mathbf{A}\cdot\mathbf{i}_\mathbf{B}=[14\cdot 6-7(3)]/7=12-3=9$

1.67 **a)** $[(x\mathbf{i}+y\mathbf{j}+z\mathbf{k})-(2\mathbf{i}-4\mathbf{j}+6\mathbf{k})]\cdot(2\mathbf{i}-4\mathbf{j}+6\mathbf{k})=0.$

$2(x-2)-4(y+4)+6(z-6)=0.\quad \therefore 2x-4y+6z=56$

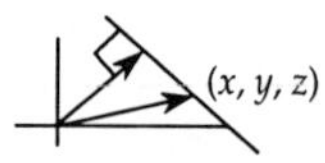

1.68 **b)** $|\mathbf{A}\times\mathbf{B}|=|15\mathbf{i}-10\mathbf{j}-24\mathbf{k}|=\sqrt{901}\cong 30$

1.69 **e)** $\nabla\phi=2x\mathbf{i}+6y\mathbf{j}-3\mathbf{k}=2\mathbf{i}+6\mathbf{j}-3\mathbf{k}.\quad |\nabla\phi|=7.\quad \therefore \mathbf{i}_\mathbf{n}=\nabla\phi/|\nabla\phi|=(2\mathbf{i}+6\mathbf{j}-3\mathbf{k})/7$

1.70 **d)** $\nabla\cdot u=2x+2y+2z=2+2+2=6$

1.71 **e)** $\nabla\times\mathbf{u}=(0-0)\mathbf{i}+(0-0)\mathbf{j}+(0-0)\mathbf{k}=0$

1.72 **a)** $\nabla\cdot\mathbf{u}=0+0+0=0.\quad \therefore$ solenoidal. $\nabla\times\mathbf{u}=(x-x)\mathbf{i}+(y-y)\mathbf{j}+(z-z)\mathbf{k}=0.$

$\therefore$ conservative and solenoidal.

Statics

by George E. Mase

Solutions to Practice Problems

2.1 **a)** $\mathbf{i}_B = \dfrac{\mathbf{i}-2\mathbf{j}-2\mathbf{k}}{\sqrt{1+4+4}} = \frac{1}{3}(\mathbf{i}-2\mathbf{j}-2\mathbf{k})$

$\mathbf{A}\cdot\mathbf{i}_B = (15\mathbf{i}-9\mathbf{j}+15\mathbf{k})\cdot\frac{1}{3}(\mathbf{i}-2\mathbf{j}-2\mathbf{k}) = 5+6-10 = 1$

2.2 **c)** $\mathbf{A}+\mathbf{B}+\mathbf{C} = (2\mathbf{i}+5\mathbf{j})+(6\mathbf{i}-7\mathbf{k})+(2\mathbf{i}-6\mathbf{j}+10\mathbf{k}) = 10\mathbf{i}-\mathbf{j}+3\mathbf{k}$

magnitude $= \sqrt{10^2+1^2+3^2} = 10.49$

2.3 **e)** $\mathbf{M} = \mathbf{r}\times\mathbf{F} = (4\mathbf{i}-6\mathbf{j}+4\mathbf{k})\times(200\mathbf{i}+400\mathbf{j})$. $M_y = 4\times 200 = 800$ since $\mathbf{k}\times\mathbf{i} = \mathbf{j}$

2.4 **c)** $\mathbf{M} = \mathbf{r}_1\times\mathbf{F}_1+\mathbf{r}_2\times\mathbf{F}_2 = (2\mathbf{i}-4\mathbf{k})\times(50\mathbf{i}-40\mathbf{k})+(-4\mathbf{i}+2\mathbf{j})\times(60\mathbf{j}+80\mathbf{k})$

$M_x = 2\times 80 = 160$ since $\mathbf{j}\times\mathbf{k} = \mathbf{i}$

2.5 **e)** Concurrent $\Rightarrow$ all pass through a point.

Coplanar $\Rightarrow$ all in the same plane.

The forces are three - dimensional.

2.6 **b)** $\sum\mathbf{F} = 0$. $\therefore \mathbf{R}+141\mathbf{i}-141\mathbf{j}-200\mathbf{i}-100\mathbf{k} = 0$

$\therefore \mathbf{R} = 59\mathbf{i}+141\mathbf{j}+100\mathbf{k}$

2.7 **c)** $\sum \mathbf{M} = 0. \quad \therefore \mathbf{M}_A + (4\mathbf{j} - 3\mathbf{k}) \times (-100\mathbf{k}) - 3\mathbf{k} \times (-200\mathbf{i}) + 4\mathbf{i} \times (141\mathbf{i} - 141\mathbf{j}) = 0$

$\therefore \mathbf{M}_A = 400\mathbf{i} - 600\mathbf{j} + 564\mathbf{k}$

2.8 **b)** They must be concurrent, otherwise a resultant moment would occur.

2.9 **a)** It is a two-force body.

2.10 **b)** $\sum M_A = 0. \quad F_B \times 8 = 400 \times 4 + 400 \times 6. \quad \therefore F_B = 500 \text{ N}$

2.11 **a)** $M_A = 400 \times 8 + 400 \times 6 = 5600 \text{ N} \cdot \text{m}$

2.12 **b)** $\sum M_B = 0. \quad 6F_A = 4 \times 300 + 600 \times 3/2. \quad \therefore F_A = 350 \text{ N}$

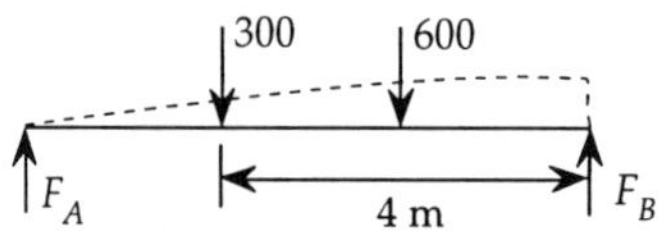

2.13 **c)** $M_A = 0.6 \times 100 - 141 \times 0.6 + 141 \times 0.8 = 88.2$

2.14 **a)** $M = 100 \sin 45° \times 4 = 282.8 \text{ cw}$

2.15 **c)** $\sum M_A = 0. \quad \therefore 6 \times 70.7 = 2 \times 0.866 F_1. \quad \therefore F_1 = 245$

$\sum F_x = 0. \quad \therefore -70.7 - 245 \times 0.5 + F_{Ax} = 0. \quad \therefore F_{Ax} = 193$

$\sum F_y = 0. \quad \therefore -70.7 - 245 \times 0.866 + F_{Ay} = 0. \quad \therefore F_{Ay} = 283$

$$\therefore F_A = \sqrt{F_{Ax}^2 + F_{Ay}^2} = \sqrt{193^2 + 283^2} = 343$$

2.16 **e)** $\sum M_A = 0. \quad \therefore 2F_B + 1.2 \times 200 - 141.4 \times 2 - 141.4 \times 1.2 + 50 = 0. \quad \therefore F_B = 81.2$

$\sum F_x = 0. \quad \therefore F_{Ax} - 200 + 141.4 = 0. \quad \therefore F_{Ax} = 58.6$

$\sum F_y = 0. \quad \therefore F_{Ay} + 81.2 - 141.4 = 0. \quad \therefore F_{Ay} = 60.2$

$$\therefore F_A = \sqrt{F_{Ax}^2 + F_{Ay}^2} = \sqrt{58.6^2 + 60.2^2} = 84.0$$

2.17 **e)** $\sum M_A = 0. \quad \therefore 500\ell + 200 \times 0.866\ell - F_C \times 2\ell = 0. \quad \therefore F_C = 337$

$0.866 F_{DC} = 337. \quad \therefore F_{DC} = 389$

$0.866 \times 389 = 0.866 F_{BD}. \quad \therefore F_{BD} = 389$

$-F_{DE} + 200 - 389 \times 0.5 - 389 \times 0.5 = 0. \quad \therefore F_{DE} = -189$

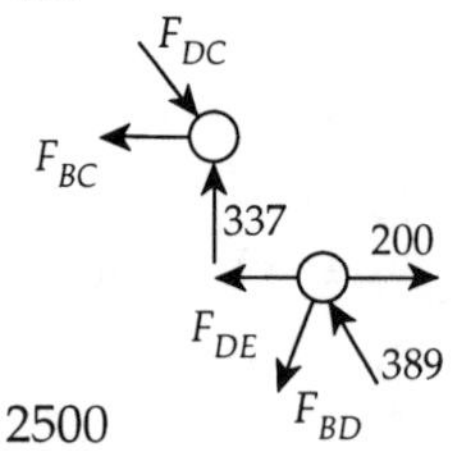

2.18 **d)** $\sum M_A = 0. \quad \therefore 5 \times 5000 = 10 \times F_C. \quad \therefore F_C = 2500 \quad \therefore F_{DC} = 2500$

$0.707 F_{BD} = 2500. \quad \therefore F_{BD} = 3536$

$0.707 \times 3536 = F_{DE}. \quad \therefore F_{DE} = 2500$

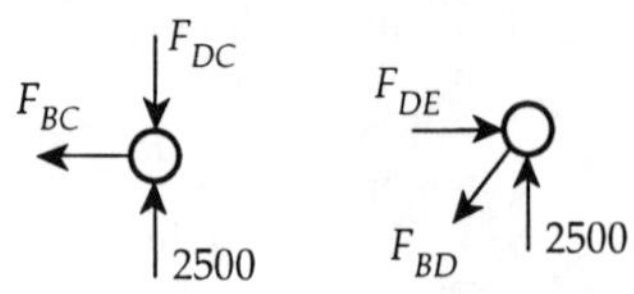

2.19 **b)** $\sum M_A = 0. \quad \therefore 4 \times 2000 + 6 \times 1000 = 8F_C. \quad \therefore F_C = 1750$

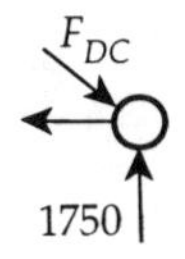

$0.707 F_{DC} = 1750. \quad \therefore F_{DC} = 2475$

Sum forces in dir. of F_{DE} : $F_{DE} - 2475 + 1000 \times 0.707 = 0.$

$\therefore F_{DE} = 1768$

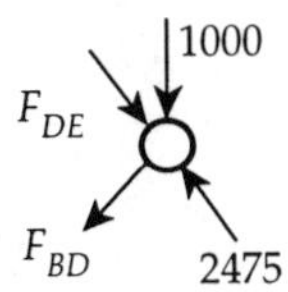

2.20 **a)** Sum forces in dir. of F_{FB} at F. $F_{FB} = 0.$

2.21 **c)** $\sum M_B = 0. \quad \therefore 12F_F = 3 \times 4000. \quad \therefore F_F = 1000 \downarrow$

$\sum F_y = 0. \quad \therefore 0.8F_{IC} = 1000. \quad \therefore F_{IC} = 1250$

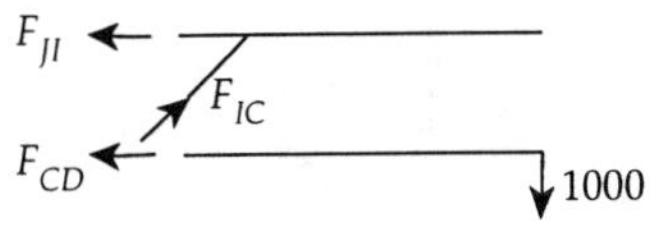

2.22 **d)** Cut vertically through link KA. Then $F_{KA} = 5000.$
Obviously, $F_{AL} = 0. \quad \therefore F_{AB} = 3000. \quad \therefore F_{BC} = 3000$

2.23 **e)** At E we see that $F_{EC} = 0. \quad \therefore$ At C, $F_{FC} = 0$

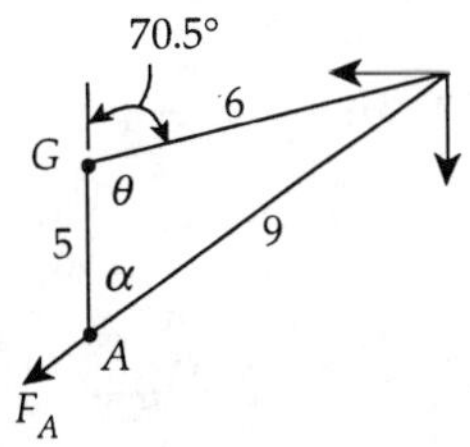

2.24 **e)** $9^2 = 6^2 + 5^2 - 2 \times 5 \times 6 \cos\theta. \quad \therefore \theta = 109.5°$

$6^2 = 9^2 + 5^2 - 2 \times 9 \times 5 \cos\alpha. \quad \therefore \alpha = 38.9°$

From pts E, C, F, B we see that $F_{EC} = F_{FC} = F_{FB} = F_{GB} = 0.$

Also, $F_A = F_{BC}$. $\sum M_G = 0.$

$\therefore 5 \times F_A \sin 38.9° + 5000 \times 6 \sin 70.5° = 5000 \times 6 \cos 70.5°.$

$\therefore F_A = -5817$

2.25 **b)** Recognize that link BC is a two-force member. $\sum M_A = 0.$

$\therefore 0.2 \times 1000 + 100 = 0.08 \times F_{BC} \times 0.8 + 0.2 \times F_{BC} \times 0.6$

$\therefore F_{BC} = 1630. \quad A_x = 1630 \times 0.8 = 1304. \quad A_y = 1630 \times 0.6 - 1000 = -22$

$\therefore F_A = \sqrt{1304^2 + 22^2} = 1304$ N

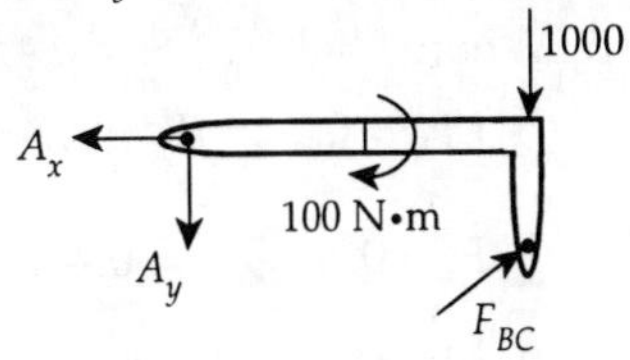

2.26 **d)** $1800 \times 0.8 = 0.6F_{BC}. \quad \therefore F_{BC} = 2400$

$0.3 \times 0.6w = 0.6 \times 2400. \quad \therefore w = 8000$

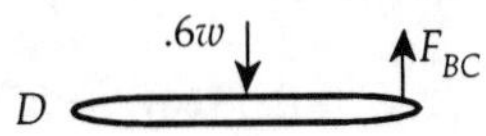

2.27 **d)** $\sum M_A = 0. \quad \therefore 1.2F_E = 0.8 \times 2400. \quad \therefore F_E = 1600$

$\therefore A_x = 2400. \quad A_y = 1600$

$\therefore F_A = \sqrt{2400^2 + 1600^2} = 2884$ N

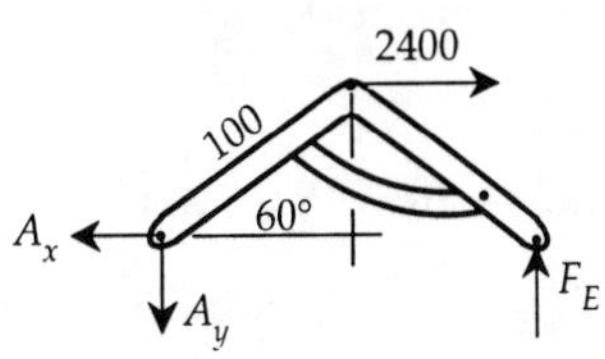

2.28 **a)** Link BD is a two-force member. $\therefore$ the force acts from D to B. Hence, the angles are found.

$120^2 = 100^2 + 100^2 - 2\times 100\times 100\cos\beta.\quad \therefore \beta = 73.7°$

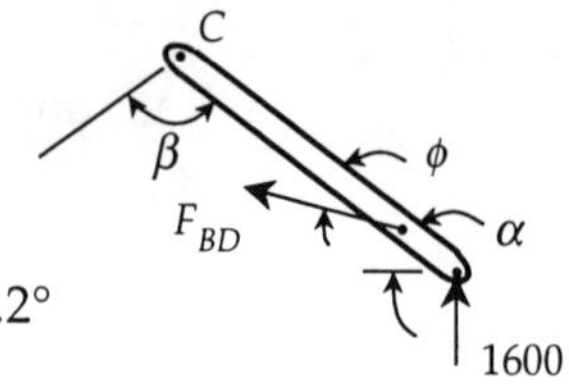

$\overline{BD}^2 = 60^2 + 40^2 - 2\times 60\times 40\cos 73.7°.\quad \therefore BD = 62.1$

$\dfrac{62.1}{\sin 73.7°} = \dfrac{40}{\sin\phi}.\quad \therefore \phi = 38.2°.\quad \alpha = (180 - 73.7)/2 = 53.2°$

$\sum M_C = 0.\quad 1600\times 100\cos 53.2° = 60\times F_{BD}\sin 38.2°.\quad \therefore F_{BD} = 2587$

2.29 **e)** $\sum F_y = 0.\quad N\times 0.866 - 980 - 0.4N\times 0.5 = 0.\quad \therefore N = 1471$

$\sum F_x = 0.\quad F = 1471\times 0.5 + 0.4\times 1471\times 0.866 = 1245$

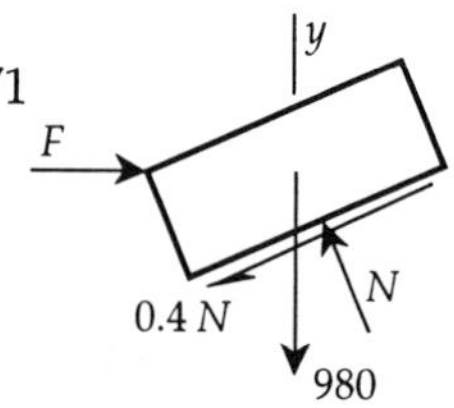

2.30 **b)** $N_1 = 490.\quad N_2 = 980.\quad \therefore F = 0.2(490 + 980) = 294$

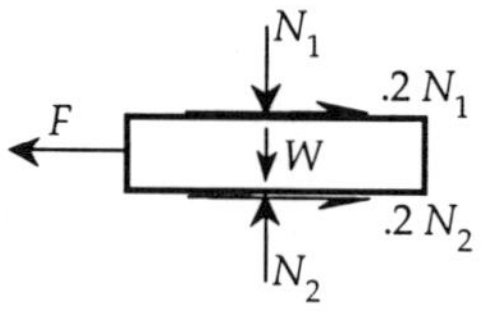

2.31 **c)** $\sum M_{\text{front wheel}} = 0.\quad \therefore 400N_2 - W\cos\theta\times 200 + W\sin\theta\times 50 = 0$

$\sum F_x = 0.\quad \therefore 0.6N_2 = W\sin\theta.$

$\therefore 400(W\sin\theta)/0.6 + 50W\sin\theta = 200W\cos\theta$

$\therefore \dfrac{\sin\theta}{\cos\theta} = \dfrac{200}{716.7} = \tan\theta.\quad \therefore \theta = 15.6°$

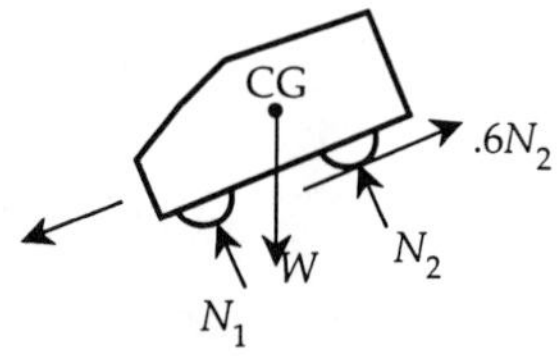

2.32 **b)** If $h < h_{\text{min}}$ then sliding occurs, and $F_f = 0.4N$.
If $h > h_{\text{min}}$ tipping occurs and $F_f < 0.4N$.
When $h = h_{\text{min}}$, $F_f = 0.4N = 0.4W = F$.
$\sum M_A = 0.\quad \therefore 4W = hF = h\times 0.4W.\quad \therefore h = 10$ cm

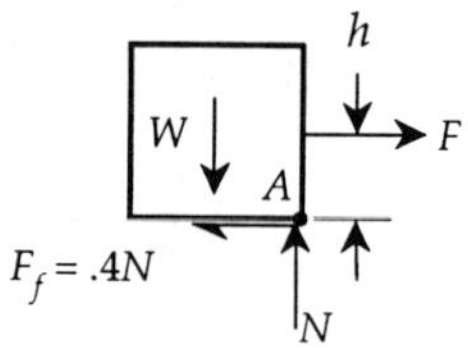

2.33 **e)** $\sum F_x = 0.\quad \therefore N_2 = 0.4N_1.$ Also, $W = 980$

$\sum M_A = 0.\quad \therefore W\cdot r = (N_1 + 0.4N_1 + 2\times 0.4N_2)r.$

$\therefore N_1 = 0.5814W = 570$

$\sum F_y = 0.\quad \therefore F = 980 - 570 - 0.16\times 570 = 319$

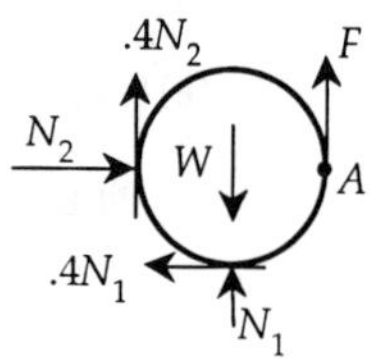

2.34 **b)** $\sum F_x = 0.\quad \therefore N_2 = 0.4N_1.\quad \sum F_y = 0.\quad \therefore N_1 + 0.4N_2 = W.$

$\therefore N_2 = 0.345W$

$\sum M_A = 0.\quad \therefore \frac{L}{2}\times W\cos\theta = N_2\times L\sin\theta + 0.4N_2\times L\cos\theta.$

This gives $\tan\theta = 1.049.\quad \therefore \theta = 46.4°$

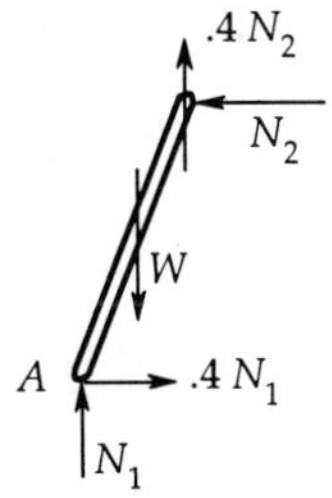

2.35 **a)** $F_B = F_D e^{-\mu\theta} = 800e^{-0.5\pi} = 166\,\text{N}$

2.36 **a)** $\sum M_A = 0. \quad \therefore 200 \times 0.6 = 0.1 \times T_1 + 0.1 \times T_2.$

$T_1 = T_2 e^{0.4\times 3\pi/2} = 6.59T_2. \quad$ Thus, $T_2 = 158$ and $T_1 = 1042.$

$\sum M_{\text{center}} = 0. \quad \therefore M = 0.1 \times (1042 - 158) = 88.4\ \text{N}\cdot\text{m}$

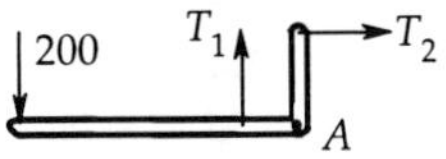

2.37 **d)** Let h = long end. m = mass/unit length. Then,
$(12 - 1.88 - h)mge^{0.5\pi} = hmg. \quad \therefore h = 8.38$ m

2.38 **d)** $\bar{x} = \dfrac{\int_0^3 xy\,dx}{\int_0^3 y\,dx} = \dfrac{\int_0^3 x^3\,dx}{\int_0^3 x^2\,dx} = \dfrac{3^4/4}{3^3/3} = 2.25$

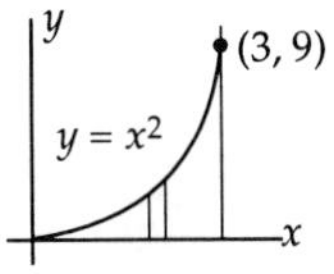

2.39 **a)** $\bar{y} = \dfrac{\int_0^3 \frac{y}{2} y\,dx}{\int_0^3 y\,dx} = \dfrac{\frac{1}{2}\int_0^3 x^4\,dx}{\int_0^3 x^2\,dx} = \dfrac{3^5/10}{3^3/3} = 2.7$

2.40 **c)** $\bar{x} = \dfrac{\int_0^1 \left(\sqrt{x} - x^2\right)x\,dx}{\int_0^1 \left(\sqrt{x} - x^2\right)dx} = \dfrac{\frac{1}{5/2} - \frac{1}{4}}{\frac{1}{3/2} - \frac{1}{3}} = 0.45$

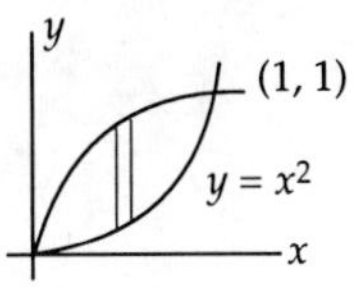

2.41 **b)** $\bar{y} = \dfrac{24 \times 3 + 6 \times 5}{6 \times 4 + 4 \times 3/2} = 3.4$

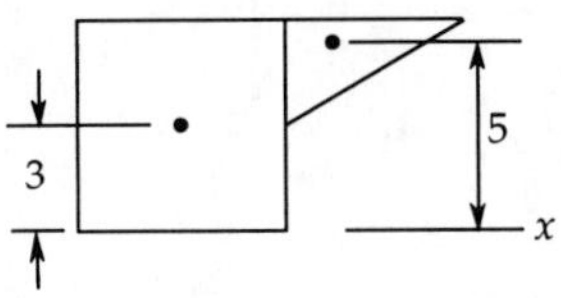

2.42 **e)** $\bar{y} = \dfrac{48 \times 3 + 12 \times 7 - \pi \times 6}{8 \times 6 + 3 \times 4 - \pi} = 3.68$

2.43 **b)** $\bar{x} = \dfrac{10 \times \frac{1}{2} + 5 \times 3.5 + 3 \times 7}{10 + 5 + 3} = 2.42$

2.44 **b)** $I_x = \int_0^3 y^3\,dx/3 = \int_0^3 x^6\,dx/3 = 3^7/21 = 104.1.$

With a horizontal strip: $I_x = \int_0^9 y^2(3 - x)dy = \int_0^9 y^2\left(3 - \sqrt{y}\right)dy = 9^3 - \dfrac{9^{7/2}}{7/2} = 104.1$

2.45 **e)** $I_x = 8 \times 6^3/3 + \left(8 \times 3^3/36 + 12 \times 7^2\right) - \left(\pi \times 1^4/4 + \pi \times 6^2\right) = 1056$

2.46 **a)** $I_y = 12\times 12^3/3 - \left(8\times 8^3/12 + 64\times 6^2\right) = 4267$. Or, alternatively:

$$I_y = 8\times 2^3/3 + 4\times 12^3/3 + 8\times 2^3/12 + 16\times 11^2 = 4267$$

2.47 **a)** $I_{\text{edge}} = I_{\text{c.g.}} + Md^2 = \frac{1}{12}M\left(b^2 + b^2\right) + M\frac{b^2}{2} = \frac{2}{3}Mb^2$

2.48 **e)** $I_x = \frac{1}{3}(6m)\times 6^2 \times 2 + 8m\times 6^2 = 432$ with $m = 1$

Dynamics

by Merle C. Potter

Solutions to Practice Problems

3.1 **d)** $v^2 = v_o^2 + 2as$. $0 = 20^2 - 2\times 5s$ $\therefore$ $s = 40$ m

3.2 **c)** $a = 5t$ $t \le 2$ $v = \int a dt = \int_0^2 5t dt + \int_2^4 10 dt = 5\frac{2^2}{2} + 10(4-2) = 30$ m/s

3.3 **c)** $v = v_o + at$. $0 = 40 - 9.8t$ $\therefore$ $t = 4.08$ s. $\therefore$ $t_{total} = 2t = 8.16$ s

3.4 **e)** $\Delta s = v\Delta t = 25 \times 0.3 = 7.5$ m

$v^2 = v_o^2 + 2as$. $0 = 25^2 - 2\times 6s$ $\therefore$ $s = 52.1$ m

$\therefore$ $s_{total} = 7.5 + 52.1 = 59.6$ m

3.5 **a)** $\theta = \alpha t^2/2 = 6\times 4^2/2 = 48$ rad. $48/2\pi = 7.64$ rev.

3.6 **e)** $\omega = \omega_o + \alpha t = 20 + 10t$. $\omega = v/L = 100/2 = 50$.

$\therefore$ $20 + 10t = 50$. $\therefore$ $t = 3$ s.

3.7 **c)** $a = \frac{v^2}{r} = \frac{20^2}{40} = 10$ m/s^2

3.8 **e)** $\frac{v^2}{r} = g. \quad \frac{v^2}{1.20} = 9.8. \quad \therefore v = 3.43 \text{ m/s}.$

$v = r\omega \quad \therefore \omega = 3.43/1.2 = 2.86 \text{ rad/s}$

3.9 **d)** $a_t = 5. \quad a_n = \frac{v^2}{R} = \frac{30^2}{200} = 4.5. \quad a = \sqrt{a_t^2 + a_n^2} = \sqrt{5^2 + 4.5^2} = 6.73 \text{ m/s}^2.$

3.10 **e)** The equation of the circle is

$s^2 + t^2 = 4. \quad 2s\frac{ds}{dt} + 2t = 0. \quad \therefore \frac{ds}{dt} = -\frac{2t}{2s} = -\frac{t}{s} = -\frac{1}{\sqrt{4-1}} = -0.577 \text{ m/s}.$

3.11 **c)** $g = \frac{V^2}{R}. \quad 9.8 = \frac{V^2}{6\,400\,000}. \quad \therefore V = 7920 \text{ m/s}.$

3.12 **a)** Because of symmetry, the time is twice that to reach the maximum height.

Let $v_y = 0: \; gt = v_0 \sin\theta. \quad \therefore t = 2\left(\frac{v_0}{g}\sin\theta\right).$

$\therefore x_{max} = v_0\left(\frac{2v_0}{g}\sin\theta\right)\cos\theta = \frac{v_0^2}{g}\sin 2\theta.$

3.13 **c)** $H = v_0^2 \sin^2\theta / 2g = 100^2 \times .707^2 2 \times 9.8 = 255 \text{ m}$

3.14 **a)** $y = v_o t \sin\theta - gt^2/2. \quad -10 = 100t \times .707 - 9.8t^2/2. \quad \therefore t = 14.6 \text{ s}$

3.15 **d)** $x = v_o t \cos\theta = 100 \times 14.6 \times 0.707 = 1032 \text{ m}$

3.16 **c)** The acceleration of the point of contact is v^2/r or $r\omega^2$. The acceleration of the center is 0.

3.17 **c)** Motion is about the point of contact. Thus,

$\omega = 10/.02 = 50 \text{ rad/s}. \quad v = r\omega = 0.4 \times 50 = 20 \text{ m/s}.$

3.18 **a)** $a = v^2/r = 20^2/0.4 = 1000 \text{ m/s}^2$

3.19 **c)** $\mathbf{a}_A = \mathbf{a}_B + \mathbf{a}_{A/B}$

$= 60\mathbf{i} - 20\mathbf{j} - .4 \times 20^2\mathbf{i} + .4 \times 100\mathbf{j} = -100\mathbf{i} + 20\mathbf{j}$

3.20 **d)** $v_B = 40\cos 45° = 28.3 = r\omega_{AB} = 1.0\omega_{AB}$

$\therefore \omega_{AB} = 28.3 \text{ ccw}.$

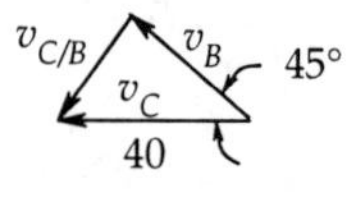

3.21 **e)** $v_{C/B} = 40\sin 45° = 28.3 = r_{BC}\omega_{BC} = 1.0\omega_{BC}. \quad \therefore \omega_{BC} = 28.3 \text{ cw}$

3.22 **a)** $\mathbf{a}_C = \mathbf{a}_B + \mathbf{a}_{C/B}$

$r_{AB}\omega_{AB}^2 = 1.0 \times 28.3^2 = 800$

$r_{BC}\omega_{BC}^2 = 1.0 \times 28.3^2 = 800$

Then, $r_{AB}\alpha_{AB} = 800. \quad \therefore \alpha_{AB} = 800 \text{ cw}$

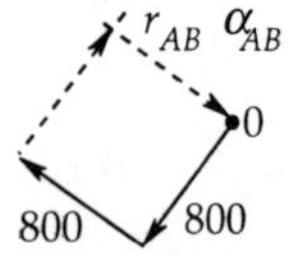

3.23 **d)** $\mathbf{a}_B = \cancel{\mathbf{a}_A}^{\,0} + \mathbf{a}_{B/A}$

$r\omega^2 = 1.0 \times 28.3^2 = 800$

$r\alpha = 1.0 \times 800 = 800$

$\therefore\ \mathbf{a}_B = -1130j$

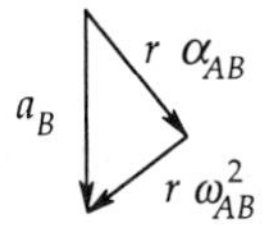

3.24 **e)** $\mathbf{v}_B = \mathbf{v}_C + \mathbf{v}_{B/C}$. $v_{B/C} = r\omega_{BC} \perp \overline{BC}$. But, $\mathbf{v}_B$ and $\mathbf{v}_C$ are both horizontal. Thus, $\mathbf{v}_{B/C} = 0$ and $\omega_{BC} = 0$.

3.25 **b)** $v_B = v_C = 20 \times 0.1 = 0.04\,\omega_{CD}$.

$\therefore\ \omega_{AB} = 50.\quad r_{CD}\omega_{CD}^2 = .1 \times 20^2 = 40$

$r_{AB}\omega_{AB}^2 = .04 \times 50^2 = 100.$

$\therefore\ r_{BC}\alpha_{BC} = \dfrac{60}{0.8} = 75.$

But, $r_{BC} = 0.1.\quad \therefore\ \alpha_{BC} = 750$ ccw.

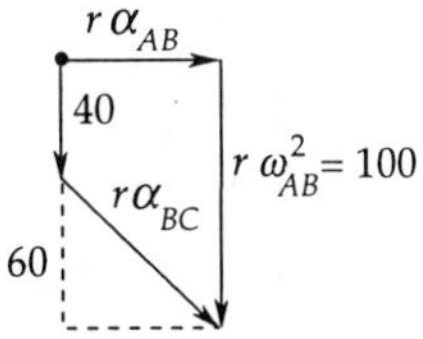

$\boldsymbol{a}_B = \boldsymbol{a}_C + \boldsymbol{a}_{B/C}$

3.26 **c)** $\mathbf{a} = -r\omega^2\mathbf{i} + 2\omega v\mathbf{j} = -0.1 \times 20^2\mathbf{i} + 2 \times 20 \times 1.0\mathbf{j} = -40\mathbf{i} + 40\mathbf{j}$

3.27 **e)** $a_{\text{Coriolis}} = 2v\omega = 2 \times 4 \times \dfrac{5 \times 2\pi}{60} = 4.19\ \text{m/s}^2.$

3.28 **c)** $a_A = 2a_B$ (from small pulley above B)

$400 \times 9.8 - 2T = 400\dfrac{a_A}{2}$ (body B)

$T - 0.2 \times 500 \times 9.8 = 500\ a_A$ (body A), $\quad \therefore\ a_A = 1.63\ \text{m/s}^2$

3.29 **a)** $W\cos 60° - W\mu \sin 60° = \dfrac{W}{g}a.$

$\therefore\ a = 9.8(0.866 - 0.3 \times 0.5) = 7.02\ \text{m/s}^2.\quad \therefore\ s = at^2/2 = 7.02 \times 10^2/2 = 351$ m

3.30 **c)** $W\sin\theta = \dfrac{W}{g}\dfrac{v^2}{r}\cos\theta \qquad \tan\theta = \dfrac{25^2}{200 \times 9.8}.\quad \therefore\ \theta = 17.7°$

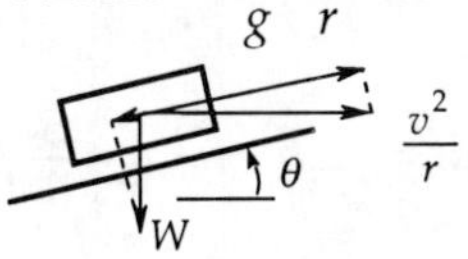

3.31 **a)** $W = \dfrac{W}{g}\dfrac{v^2}{r}.\quad \therefore\ v = \sqrt{9.2 \times 6\,600\,000} = 7790\ \text{m/s}$

3.32 **c)** $F = k\dfrac{m_1 m_2}{r^2}.\quad W = k\dfrac{m_e W/g}{r^2}.$ (Let W be your weight!)

$\therefore\ m_e = r^2 g/k = 6\,400\,000^2 \times 9.8/6.67 \times 10^{-11} = 6 \times 10^{24}$ kg

3.33 **d)** $v^2 = v_0^2 + 2as.\quad -0.6W = \dfrac{W}{g}a.\quad \therefore\ a = -.06g.\quad 0 = 25^2 - 2 \times .6 \times 9.8s.$

$\therefore\ s = 53.2$ m

3.34 **e)** $\Sigma M = 0$

For maximum accel. the force on the front wheel = 0.

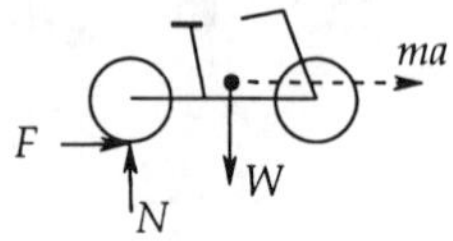

Thus, $30W = 80\ ma$. $\therefore a = \frac{30 \times 9.8}{80} = 3.68\ \text{m/s}^2$

3.35 **b)** Before: $\Sigma F_y = 0$. $T\cos 30° = W$. $\therefore T = \frac{W}{0.866}$

After: $F = 0$, $\Sigma F_n = ma_n$. $n \perp t$. $\therefore$ $T = W\cos$ $= 0.866\ W$.

ratio $= \frac{\text{before}}{\text{after}} = \frac{1}{0.866 \times 0.866} = 1.33$

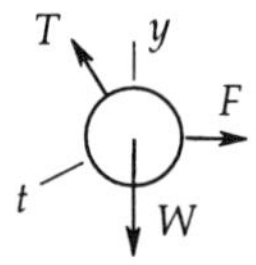

3.36 **d)** Take moments about the back wheels.

$4W - 8N_2 = 1.2\frac{W}{g} \times 2$. $\therefore N_2 = 0.469W = .469 \times 8000 \times 9.8 = 36\ 800\ \text{N}$

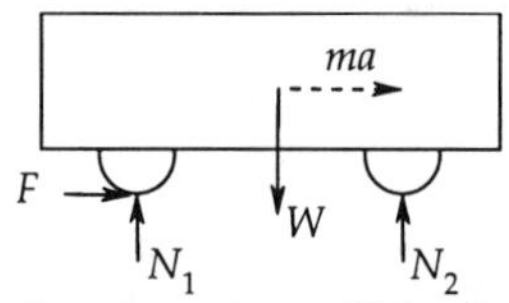

3.37 **b)** Acceleration of block = 0.2α. Thus, $T = 50 \times 0.2\alpha + 50 \times 9.8 \times 0.2$.

$\Sigma M = I\alpha$.

Use $I = mk^2$. $(40 \times 9.8 - 40\alpha \times .1) \times .1 - (50 \times .2\alpha + 50 \times 9.8 \times .2) \times .2 = 30 \times .1^2\alpha$.

$\therefore \alpha = 7.26\ \text{rad/s}^2$

3.38 **c)** $\Sigma M_o = I_o\alpha$. $mg \times \frac{\ell}{2} = \frac{1}{3}m\ell^2\alpha$. $\therefore \alpha = \frac{3g}{2\ell}$.

$\Sigma F_y = m\bar{a}$ ($\bar{a}$ is acc. of mass center).

$mg - F_o = m\frac{\ell}{2}\alpha = m\frac{3}{4}g$. $\therefore F_o = \frac{mg}{4}$.

3.39 **d)** $F = 5t$ for $t < 2$ s. $\therefore 5t = 2\frac{dv}{dt}$ or $\int_0^t 5t dt = \int_0^v 2dv$. $\therefore v = \frac{5}{4}t^2$.

$v = \frac{dx}{dt} = \frac{5}{4}t^2$. $\therefore \int_0^x dx = \frac{5}{4}\int_0^2 t^2 dt$ and $x = \frac{5}{4} \times \frac{2^3}{3} = \frac{10}{3}$ m.

3.40 **e)** $v = \frac{5}{4}t^2 = \frac{5 \times 2^2}{4} = 5$ m/s after 2 seconds. For $t > 2$, $20 - 5t = 2\frac{dv}{dt}$.

$\therefore \int_2^4 (20 - 5t)dt = 2\int_5^v dv$. $20 \times (4 - 2) - \frac{5}{2}(16 - 4) = 2(v - 5)$. $\therefore v = 10$ m/s.

3.41 **c)** From Problem 3.40,

$$\int_2^t (20-5t)dt = 2\int_5^v dv \quad \text{or} \quad 20(t-2)-\frac{5}{2}\left(t^2-4\right)=2(v-5).$$

The mass reaches its maximum displacement when $v = 0$. Set $v = 0$ and solve for t:

$$20t-40-\frac{5}{2}t^2+10=-10. \quad \therefore t^2-8t+8=0. \quad \therefore t=\frac{8\pm\sqrt{64-32}}{2}=6.83 \text{ s}.$$

3.42 **b)** $W_{net}=\Delta E=\frac{1}{2}mv^2. \quad 60\times10-\frac{1}{2}\times100\times2^2=\frac{1}{2}50v^2. \quad \therefore v=4 \text{ m/s}$

3.43 **d)** $W_{net}=\Delta KE=\frac{1}{2}mv^2+\frac{1}{2}I\omega^2. \quad v=r\omega.$

$$\therefore\ 100\times4=\frac{1}{2}\times100(0.2\omega)^2+\frac{1}{2}\left(\frac{1}{2}\times100\times.2^2\right)\omega^2. \quad \therefore \omega=11.55 \text{ rad/s}$$

3.44 **d)** $W_{net}=\Delta KE=\frac{1}{2}mv^2. \quad \Delta x=50-30=20.$

$$\therefore\ 2\times9.8\times0.4-\frac{1}{2}\times100\times.2^2=\frac{1}{2}\times2v^2. \quad \therefore v=2.42 \text{ m/s}$$

3.45 **e)** $W_{net}=\Delta KE=\frac{1}{2}mv^2+\frac{1}{2}I\omega^2.$

$$\frac{1}{2}100\left(.4^2-.1^2\right)-10\times9.8(.25-.20)=\frac{1}{2}10\left(\frac{v}{2}\right)^2+\frac{1}{2}\left(10\times.5^2/12\right)\left(\frac{v}{.5}\right)^2$$

where $\overline{v}=v/2$ and $\omega=v/.5$. The above gives $v=1.25$ m/s.

3.46 **b)** $\Sigma M_C\Delta t=I_C\omega$ where C is the point of contact. $I_C=I_o+mr^2=\frac{3}{2}mr^2.$

$$100\times0.4\times4=\frac{3}{2}\times100\times.2^2\omega. \quad \therefore \omega=26.7 \text{ rad/s}.$$

3.47 **a)** Integrate Eq. 3.2.1 as follows: $\int Fdt=m\Delta v. \quad F=400t.$

$$F_f=\mu N=0.2\times100\times0.866\times9.8=170$$

$$400t=100\times0.5\times9.8+170. \quad \therefore t=1.65 \text{ s when motion initiates}.$$

$$\int_{1.65}^{2.0} 400tdt-(170+50\times9.8)\times0.35=100v. \quad \therefore v=0.245 \text{ m/s}$$

3.48 **e)** $10\,m-20\,m = v_A'm+v_B'm. \quad \therefore\ v_A'+v_B'=-10$

$0.8=\frac{v_B'-v_A'}{20+10}. \quad \therefore v_B'-v_A'=24.$ Simultaneous solutions yields $v_B'=7$ m/s, $v_A'=-17$ m/s.

3.49 **c)** $\Delta KE=\frac{1}{2}mv_A^2+\frac{1}{2}mv_B^2-\frac{1}{2}mv_A'^2-\frac{1}{2}mv_B'^2=\frac{2}{2}\left(10^2+20^2\right)-\frac{2}{2}\left(7^2+17^2\right)=162 \text{ J}.$

Mechanics of Materials

by George E. Mase

Solutions to Practice Problems

Metric Units

4.1 **b)** Isotropic

4.2 **e)** Poisson's Ratio

4.3 **d)**

4.4 **d)** $\sigma = E\delta/L = P/\pi r^2$

$\therefore P = E\,\delta\pi r^2/L = 210\times10^9\times0.001\times\pi\times0.01^2/1 = 66\,000\text{ N}$

4.5 **d)** $\dfrac{P}{\pi r^2} = E\dfrac{\delta}{L}. \quad \therefore \delta = \dfrac{14\,000\times30}{\pi\times0.01^2\times210\times10^9} = 0.0064\text{ m}$

4.6 **a)** $\dfrac{P}{A} = E\dfrac{\delta}{L}. \quad \therefore P = \dfrac{AE\delta}{L} = 0.025\times0.003\times\left(210\times10^9\right)\times\dfrac{0.0008}{1} = 12\,600\text{ N}$

4.7 **b)** $\dfrac{P}{A} = G\dfrac{\delta}{L}. \quad \therefore G = \dfrac{PL}{A\delta} = \dfrac{20000\times0.15}{0.012\times0.2\times0.00003} = 41.7\times10^9\text{ Pa}$

4.8 **a)** $P/A = E\varepsilon. \quad \varepsilon = \dfrac{600\,000}{\pi\times0.025^2\times210\times10^9} = 0.001455$

$\Delta d = \nu\varepsilon d = 0.28\times0.001455\times5 = 0.00204\text{ cm}$

$d = d - \Delta d = 5 - 0.00204 = 4.9980\text{ cm}$

4.9 **b)** $P/A = E\varepsilon. \quad \varepsilon = \dfrac{1500\,000}{\pi \times 0.06^2 \times 70 \times 10^9} = 0.00190$

$$\Delta d = \nu \varepsilon d = 0.33 \times 0.0019 \times 12.015 = 0.0075 \text{ cm}$$

$$d = d - \Delta d = 12.015 - 0.0075 = 12.008 \text{ cm}$$

4.10 **e)** $\sigma = E\delta/L. \quad 100 \times 10^6 = 210 \times 10^9 (\delta/L). \quad \therefore \ \delta = 4.76 \times 10^{-4} L$

$$\delta_T = \alpha L \Delta T = 11.7 \times 10^{-6} \times 30L = 3.51 \times 10^{-4} L. \quad \delta_{final} = \delta - \delta_T = 1.25 \times 10^{-4} L$$

$$\therefore \sigma = \left(210 \times 10^9\right) \times \left(1.25 \times 10^{-4}\right) = 26.2 \times 10^6 \text{ Pa}$$

4.11 **a)** $\delta = \alpha L \Delta T = \left(11.7 \times 10^{-6}\right) \times 300 \times 75 = 0.263 \text{ m}$

4.12 **d)** $\dfrac{\delta}{L} = \dfrac{\sigma}{E} = \alpha \Delta T. \quad \dfrac{70 \times 10^6}{70 \times 10^9} = 23 \times 10^{-6} (30 - T). \quad \therefore T = -13.5°\text{C}$

4.13 **d)** It expands at a different rate.

4.14 **a)** $\tau = \dfrac{Tr}{J} = \dfrac{200 \times 0.03}{\pi \times 0.06^4 / 32} = 4.72 \times 10^6 \text{ Pa}$

4.15 **e)** $J = \pi\left(a_1^4 - a_2^4\right)/2 = \pi\left(0.03^4 - .025^4\right)/2 = 65.9 \times 10^{-8}$

$$\tau = \frac{Tr}{J} = \frac{200 \times 0.03}{65.9 \times 10^{-8}} = 9.10 \times 10^6 \text{ Pa}$$

4.16 **a)** $T = \dfrac{\tau J}{r} = \dfrac{\left(140 \times 10^6\right) \times \pi \times .05^4/2}{0.05} = 27\,500 \text{ N} \cdot \text{m}$

4.17 **d)** $\theta = \dfrac{TL}{JG} = \dfrac{200 \times 0.3 \times 0.5}{\left(83 \times 10^9\right) \times \pi \times .05^4/2} = 0.368 \text{ rad or } 21.1°$

4.18 **d)** $\tau_{\max} = \dfrac{Tr}{J} = \dfrac{1200 \times 0.04}{\pi \times 0.04^4/2} = 11.9 \times 10^6 \text{ Pa}.$

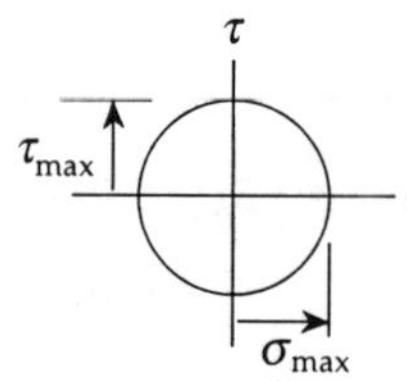

$$\therefore \sigma_{\max} = 11.9 \times 10^6 \text{ Pa}.$$

4.19 **b)** $\sigma = My/I. \quad \sigma_{\max}$ occurs at $y = y_{\max}$

4.20 **a)** A triangle.

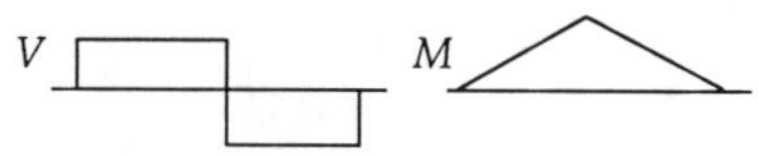

4.21 **a)** $\sum M_{right} = 0. \quad \therefore 8F_{left} = 4000 \times 4 + 4000 \times 2. \quad \therefore F = 3000.$

$$M_A = 3000 \times 4 = 12000 \text{ N} \cdot \text{m}$$

4.22 **b)** $M_A = 4000 \times 4 + 4000 \times 2 = 24000 \text{ N} \cdot \text{m}$

4.23 **b)** $4F_{right} = 24\,000 \times 2 + 4\,000 \times 6$. $\therefore F_{right} = 18\,000$ N. $4F_{left} = 24\,000 \times 2 - 4\,000 \times 2$.

$\therefore F_{left} = 10\,000$ N. $M_{\max} =$ area under diagram $= 10\,000 \times 1.667/2 = 8\,330$ N·m.

$$\sigma = \frac{My}{I} = \frac{8\,330 \times 0.1}{971 \times 10^{-8}} = 85.8 \times 10^6 \text{ Pa}$$

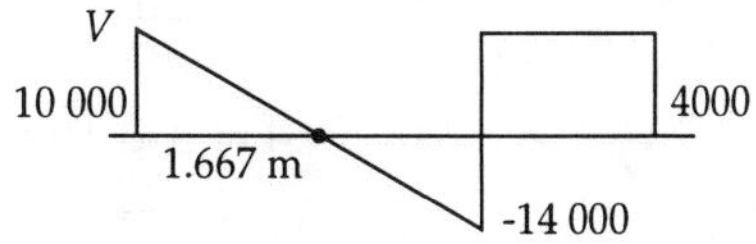

4.24 **c)** Compression occurs in bottom fibers over right support.

There, $M = 4\,000 \times 2 = 8\,000$ N·m. $\sigma = \dfrac{My}{I} = \dfrac{8\,000 \times 0.1}{971 \times 10^{-8}} = 82.4 \times 10^6$ Pa

4.25 **e)** $\tau_{\max} = \dfrac{VQ}{Ib} = \dfrac{14\,000(0.002 \times 0.05)}{\left(971 \times 10^{-8}\right) \times 0.02} = 7.21 \times 10^6$ Pa

4.26 **c)** $\tau_{\max}$ occurs on the N.A. with a sudden decrease when b goes from 2 to 16 cm.

Also, it is a parabolic distribution.

4.27 **c)** $\sigma = \dfrac{My}{I}$. $\dfrac{I}{y} = \dfrac{M}{\sigma} = \dfrac{24\,000}{140 \times 10^6} = 171 \times 10^{-6}$ m^3

4.28 **b)** Using the area under the curve: $M_{\max} = 1000 \times 3 + 3000 \times 3/2 = 7500$ N·m

$$\sigma_{\max} = \frac{My}{I} = \frac{7500 \times 0.025}{0.1 \times 0.05^3/12} = 180 \times 10^6 \text{ Pa}$$

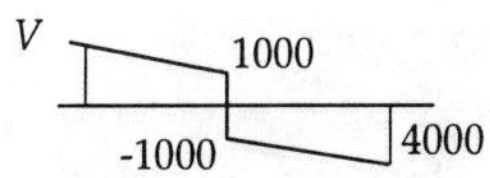

4.29 **e)** $\sigma_{\max} = \dfrac{My}{I} = \dfrac{7500 \times 0.05}{0.05 \times 0.1^3/12} = 90 \times 10^6$ Pa

4.30 **a)** $V_{\max} = 1000$. $\tau_{\max} = \dfrac{VQ}{Ib} = \dfrac{1000(0.025 \times 0.05 \times 0.0125)}{\left(0.05 \times 0.05^3/12\right) \times 0.05} = 600 \times 10^3$ Pa

4.31 **d)** $\delta = \dfrac{PL^3}{48EI} = \dfrac{2000 \times 6^3}{48 \times \left(210 \times 10^9\right) \times 0.05^4/12} = 0.0823$ m

4.32 **e)** $\delta = \dfrac{PL^3}{48EI} + \dfrac{5wL^4}{384EI}$. $I = \dfrac{bh^3}{12} = \dfrac{0.1 \times 0.05^3}{12} = 1.04 \times 10^{-6}$

$$= \frac{2000 \times 6^3}{48 \times \left(210 \times 10^9\right) \times 1.04 \times 10^{-6}} + \frac{1000 \times 5 \times 6^4}{384 \times \left(210 \times 10^9\right) \times 1.04 \times 10^{-6}} = 0.118 \text{ m}$$

4.33 **a)** $\delta = \theta L_2 = \dfrac{PL^2}{16EI} \times 4$. $0.1 = \dfrac{P \times 8^2 \times 4}{16 \times \left(210 \times 10^9\right)\pi \times .025^4/4}$. $\therefore P = 403$ N

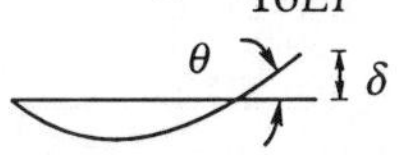

4.34 **d)** $\tau_{\max} = \dfrac{1}{2}\sqrt{\left(\sigma_x - \sigma_y\right)^2 + 4\tau^2} = \dfrac{1}{2}\sqrt{60^2 + 4 \times 40^2} = 50$ MPa

4.35 **a)** $\alpha_{\max} = \dfrac{1}{2}\left(\sigma_x + \sigma_y\right) + \tau_{\max} = 0 + 40 = 40$ MPa

4.36 **c)** $\tau_{max} = \frac{1}{2}\sqrt{(30+50)^2 + 4\times 30^2} = 50 \text{ MPa}$

4.37 **d)** $\tau = Tr/J = 600\times 0.025 \Big/ \frac{\pi\times .025^4}{2} = 24.45\times 10^6 \text{ Pa}$

$\sigma = P/A = 40\,000 \big/ \pi\times 0.025^2 = 20.37\times 10^6 \text{ Pa}$

$\therefore \tau_{max} = \frac{1}{2}\sqrt{20.37^2 + 4\times 24.45^2} = 26.5\times 10^6 \text{ Pa}$

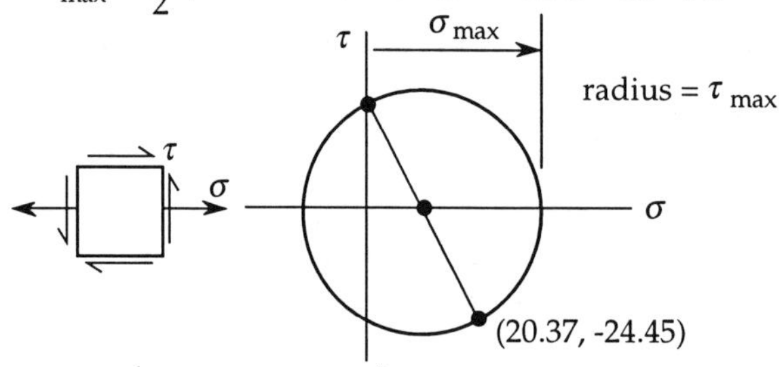

4.38 **c)** $\sigma_{max} = \left(\frac{1}{2}\times 20.37 + 26.5\right)\times 10^6 = 36.7\times 10^6 \text{ Pa}$

4.39 **b)** $My/I = 3\,200\times 0.025 \Big/ \frac{\pi\times .025^4}{4} = 261\times 10^6$ comp.

$P/A = 40\,000 \big/ \pi\times 0.025^2 = 20.4\times 10^6$ Pa tension.

$\sigma_A = (261 - 20.4)\times 10^6 = 241\times 10^6 \text{ Pa}$

4.40 **c)** $\tau_{max} = \sigma/2 = 120\times 10^6$ Pa. $VQ/Ib = 0$ on outer fibers.

4.41 **c)** $\tau = \frac{Tr}{J} = \frac{8\,000\times 0.25\times 0.025}{\pi\times .025^4/2} = 81.5\times 10^6 \text{ Pa}$

$\sigma = \frac{My}{I} = \frac{8\,000\times 0.4\times 0.025}{\pi\times .025^4/4} = 261\times 10^6 \text{ Pa}$

$\tau_{max} = \frac{1}{2}\sqrt{261^2 + 4\times 81.5^2}\times 10^6 = 154\times 10^6 \text{ Pa}$

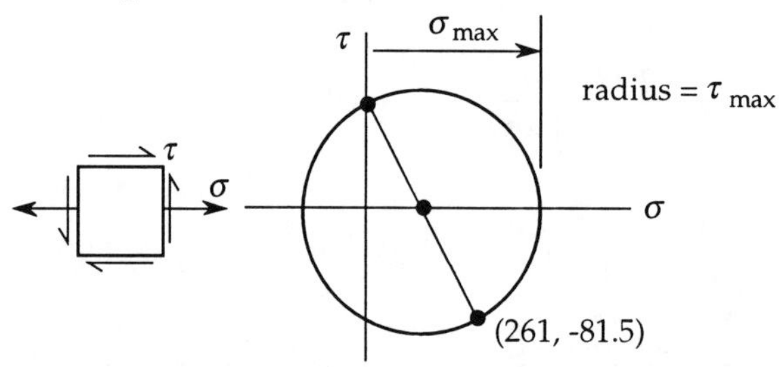

4.42 **a)** $\sigma_{max} = \left(\frac{1}{2}\times 261 + 154\right)\times 10^6 = 284\times 10^6 \text{ Pa}$

4.43 **c)** $M_{max} = FL_2 = 2000\times 1 = 2000 \text{ N}\cdot\text{m}.$

$\sigma = \frac{My}{I} = \frac{2000\times 0.025}{0.02\times 0.05^3/12} = 240\times 10^6 \text{ Pa}.$

4.44 **d)** $\frac{VQ}{Ib} = \frac{2000 \times (0.02 \times 0.025) \times 0.0125}{\left(0.02 \times 0.05^3/12\right) \times 0.02} = 3 \times 10^6$ Pa.

4.45 **a)** The force F provides a torque of 2000 N·m and a force of 2000 N acting on the end of the steel shaft. This is best described by (a) since the top fibers experience tension. There is no normal stress in the circumferential direction, ruling out (b), (c) and (d).

4.46 **e)** $\sigma = \frac{My}{I} = \frac{(2000 \times 2) \times 0.025}{\pi \times 0.025^4/4} = 326 \times 10^6$ Pa.

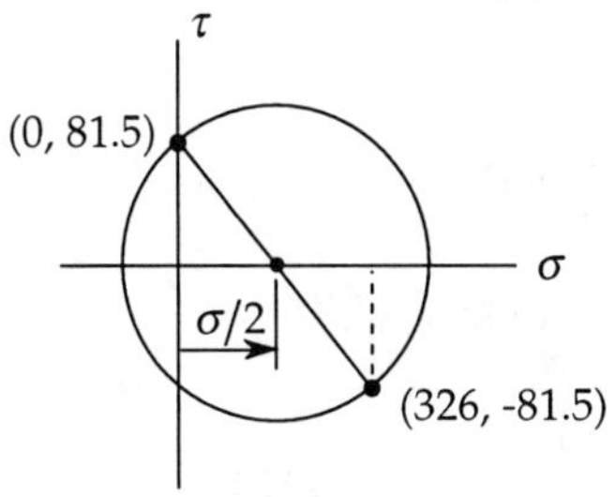

$\tau = \frac{Tr}{J} = \frac{(2000 \times 1) \times 0.025}{\pi \times 0.025^4/2} = 81.5 \times 10^6$ Pa.

$\therefore \sigma_{max} = 163 + \left[163^2 + 81.5^2\right]^{1/2}$

$= 345$ MPa.

4.47 **d)** $\delta = \frac{PL^3}{3EI} = \frac{2000 \times 2^3}{3 \times 210 \times 10^9 \times \pi \times 0.025^4/4} = 0.0828$ m.

4.48 **b)** $\sigma_t = pD/2t. \quad \therefore p = 180 \times 10^6 \times 2 \times 0.005/0.8 = 2250 \times 10^3$ Pa

4.49 **c)** $\sigma_a = pD/4t. \quad \therefore t = \frac{8000 \times 10^3 \times 1.2}{4 \times 200 \times 10^6} = 0.012$ m

4.50 **e)** $\tau_{max} = \frac{1}{2}\sqrt{(200-200) + 0 \times 4} = 0$

4.51 **a)** $\left(\frac{\Delta L}{L}\right)_s = \left(\frac{\Delta L}{L}\right)_c. \quad \therefore \varepsilon_s = \varepsilon_c. \quad \therefore \sigma_s = \frac{E_s}{E_c}\sigma_c = 10.5\,\sigma_c.$

$F_s + F_c = 2000000$ or $A_s\sigma_s + A_c\sigma_c = 2000000.$

$\sigma_s\left[\pi\left(0.137^2 - 0.125^2\right) + \pi \times \frac{0.125^2}{10.5}\right] = 2 \times 10^6. \quad \therefore \sigma_s = 137 \times 10^6$ Pa.

4.52 **d)** $n = E/E_{min} = 3.$ The area is transformed:

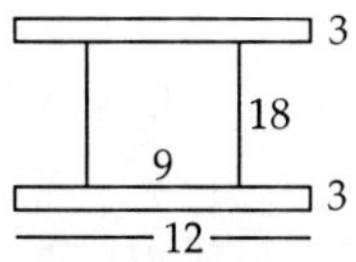

$I_t = \frac{0.12 \times 0.24^3}{12} - \frac{0.03 \times 0.18^3}{12} = 1.237 \times 10^{-4}$ m^4.

$\therefore \sigma_{al} = \frac{My}{I} = \frac{2000 \times 2 \times 0.12}{1.237 \times 10^{-4}} = 3.88 \times 10^6$ Pa, $\sigma_s = \frac{nMy}{I} = \frac{3 \times 2000 \times 2 \times 0.09}{1.237 \times 10^{-4}} = 4.37 \times 10^6$ Pa

4.53 **c)** $n = E/E_{\min} = 3$. The area is transformed:

$$I_t = \frac{0.36\times 0.24^3}{12} - \frac{0.33\times 0.18^3}{12} = 2.54\times 10^{-4}\ \text{m}^4.$$

$$\therefore \sigma_s = \frac{nMy}{I} = \frac{3\times 4000\times 0.12}{2.54\times 10^{-4}} = 5.67\times 10^6\ \text{Pa}$$

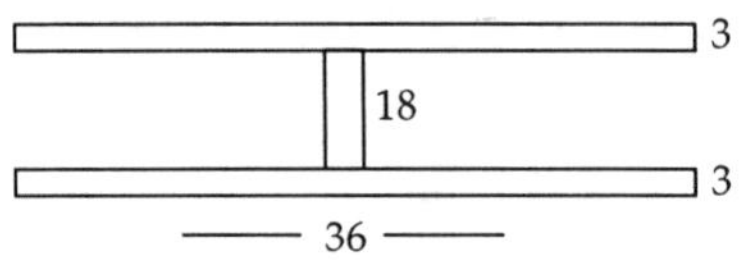

4.54 **d)** $60 = \frac{L}{r} = \frac{L}{\sqrt{I/A}} = \frac{L}{\sqrt{0.1\times\left(0.1^3/12\right)/0.01}}$. $\therefore L = 1.73$ m

4.55 **c)** Assume $P = 4000$ N using a factor of safety of 2.

$$4000 = \frac{\pi^2\times 70\times 10^9\times \pi\times\left(0.1^4/64\right)}{4L^2}. \quad \therefore L = 14.55\ \text{m}$$

4.56 **e)** $P_{cr} = 4\pi^2 EI/L^2 = \alpha\Delta TEA$. $\Delta T = \frac{4\pi^2 I}{\alpha AL^2} = \frac{4\pi^2\times\pi\times\left(0.02^4/64\right)}{11.7\times 10^{-6}\times\pi\times 0.01^2\times 4^2} = 5.27\ ^\circ\text{C}$

4.57 **b)** $P_{cr} = 4\pi^2 EI/L^2 = 30\,000$. $\therefore \pi^2 EI/L^2 = 7500$. $\therefore P_{cr} = \pi^2 EI/4L^2 = 7500/4 = 1875$ N

Solutions to Practice Problems

English Units

4.1 **b)** Isotropic

4.2 **e)** Poisson's Ratio

4.3 **d)**

4.4 **d)** $\sigma = E\delta/L = P/\pi r^2$. $\therefore P = E\delta\pi r^2/L = 30\times 10^6\times 0.04\times\pi(1/4)^2/48 = 4909$ lb.

4.5 **d)** $\frac{P}{\pi r^2} = E\frac{\delta}{L}$. $\therefore \delta = \frac{PL}{\pi r^2 E} = \frac{3500\times 100\times 12}{\pi\times\left(1/4^2\right)\times 30\times 10^6} = 0.71$ in

4.6 **a)** $\frac{P}{A} = E\frac{\delta}{L}$. $\therefore P = \frac{AE\delta}{L} = \frac{1}{8}\times 1\times 30\times 10^6\times\frac{1/32}{36} = 3255$ lb.

4.7 **b)** $\frac{P}{A} = G\frac{\delta}{L}$. $\therefore G = \frac{PL}{A\delta} = \frac{5000\times 6}{8\times(1/2)\times 0.0012} = 6.25\times 10^6$

4.8 **a)** $\frac{P}{A} = E\varepsilon. \quad \therefore \varepsilon = \frac{150,000}{\pi \times 1^2 \times 30 \times 10^6} = 0.00159$

$\therefore \Delta d = \nu \varepsilon d = 0.28 \times 0.00159 \times 2 = 0.00089. \quad \therefore d_{after} = 2 - 0.00095 = 1.9991 \text{ in}$

4.9 **b)** $\frac{P}{A} = E\varepsilon. \quad \therefore \varepsilon = \frac{400,000}{\pi \times (5.923/2)^2 \times 10 \times 10^6} = 0.00145$

$\Delta d = \nu \varepsilon d = 0.33 \times 0.00145 \times 5.923 = 0.00283.$

$\therefore d_f = d - \Delta d = 5.923 - 0.00283 = 5.920 \text{ in}$

4.10 **e)** $\sigma = E\delta/L. \quad 16,000 = 30 \times 10^6 (\delta/L). \quad \therefore \delta = 5.33 \times 10^{-4} L.$

$\delta_T = \alpha L \Delta T = 6.5 \times 10^{-6} \times 50L = 3.25 \times 10^{-4} L.$

$\therefore \delta_{final} = 2.08 \times 10^{-4} L. \quad \therefore \sigma = 30 \times 10^6 \times 2.08 \times 10^{-4} = 6,240 \text{ psi}$

4.11 **a)** $\delta = \alpha L \Delta T = 6.5 \times 10^{-6} \times 1000 \times 12 \times 130 = 10.1"$

4.12 **d)** $\frac{\delta}{L} = \frac{\sigma}{E} = \alpha \Delta T. \quad \therefore \frac{10,000}{10 \times 10^6} = 12.8 \times 10^{-6}(80 - T). \quad \therefore T = 1.88°\text{F}$

4.13 **d)** It expands at a different rate.

4.14 **a)** $\tau = \frac{Tr}{J} = \frac{2000 \times 1}{\pi \times 1^4/2} = 1,273 \text{ psi}$

4.15 **e)** $J = \pi\left(a_1^4 - a_2^4\right)/2 = \pi\left(1^4 - .875^4\right)/2 = 0.650 \text{ in}^4$

$\tau = \frac{Tr}{J} = 2,000 \times 1/0.650 = 3,077 \text{ psi}$

4.16 **a)** $T = \frac{\tau J}{r} = 20,000 \times \left(\pi 2^4/2\right)/2 = 251,300 \text{ in} \cdot \text{lb or } 20,940 \text{ ft} \cdot \text{lb}$

4.17 **d)** $\theta = \frac{TL}{JG} = \frac{160 \times 24 \times 18}{\left[\pi (7/16)^4/2\right] \times 12 \times 10^6} = 0.1001 \text{ rad } (5.73°)$

4.18 **d)** $\tau_{max} = \frac{Tr}{J} = \frac{(1200 \times 12) \times 2}{\pi \times 2^4/2} = 1146 \text{ psi}. \quad \therefore \sigma_{max} = 1146 \text{ psi}.$

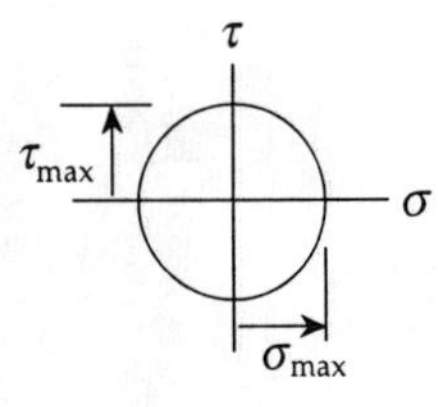

4.19 **b)** $\sigma = My/I. \quad \therefore \sigma = \sigma_{max}$ at $y = y_{max}$

4.20 **a)** A triangle.

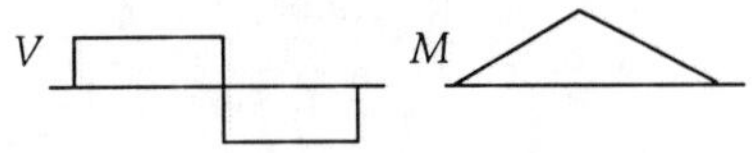

4.21 **a)** $\sum M_{right\ end} = 0. \quad \therefore 20F_{left} = 1,000 \times 10 + 1,000 \times 5. \quad \therefore F_{left} = 750.$

$M_A = 750 \times 10 = 7,500 \text{ lb} \cdot \text{ft}$

4.22 **b)** $M_A = 1,000 \times 10 + 1,000 \times 5 = 15,000 \text{ ft} \cdot \text{lb}$

4.23 a) $10F_{right} = 5{,}000 \times 5 + 1{,}000 \times 14. \quad \therefore F_{right} = 3{,}900 \text{ lb.}$

$10F_{left} = 5{,}000 \times 5 - 1{,}000 \times 4. \quad \therefore F_{left} = 2{,}100 \text{ lb.}$

$M_{max} = 2{,}100 \times (4.2/2) = 4{,}410 \text{ ft} \cdot \text{lb} = \text{area under V - diagram.}$

$$\sigma = \frac{My}{I} = \frac{4{,}410 \times 12 \times 5}{60.67} = 4{,}361 \text{ psi}$$

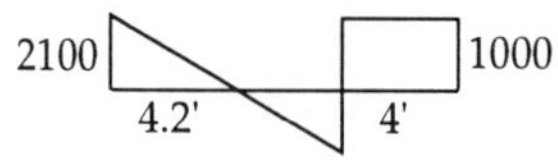

4.24 b) Compression occurs in bottom fibers over right support.

$$\sigma = \frac{My}{I} = \frac{4{,}000 \times 12 \times 5}{60.67} = 3{,}956 \text{ psi}$$

4.25 e) $\tau_{max} = \dfrac{VQ}{Ib} = \dfrac{2{,}900 \times (5 \times 2.5)}{60.67 \times 1} = 597 \text{ psi}$

4.26 c) τ_{max} occurs on the N.A. with a sudden decrease when b goes from 1" to 8".

Also, it is parabolic.

4.27 c) $\sigma = \dfrac{My}{I}. \quad \therefore \dfrac{I}{y} = \dfrac{M}{\sigma} = \dfrac{15{,}000 \times 12}{2{,}000} = 9 \text{ in}^3$

4.28 b) Area under curve: $M_{max} = 250 \times 8 + 800 \times (8/2) = 5{,}200$

$$\sigma = \frac{My}{I} = \frac{5{,}200 \times 12 \times 1}{4 \times 2^3/12} = 23{,}4000 \text{ psi}$$

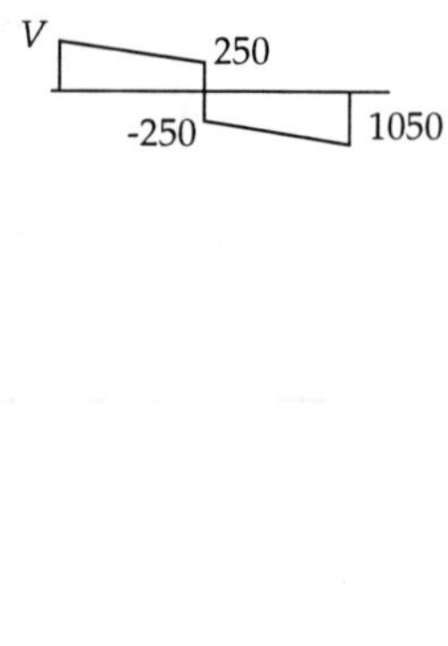

4.29 e) $\sigma_{max} = \dfrac{5{,}200 \times 12 \times 2}{2 \times 4^3/12} = 11{,}700 \text{ psi}$

4.30 e) $V_{max} = 250. \quad \tau_{max} = \dfrac{VQ}{Ib} = \dfrac{250 \times 2 \times (1/2)}{2 \times (2^3/12) \times 2} = 93.8$

4.31 d) $\delta = PL^3/48EI = 500 \times 240^3/(48 \times 30 \times 10^6 \times 2^4/12) = 3.6"$

4.32 e) $\delta = \dfrac{PL^3}{48EI} + \dfrac{5wL^4}{384EI} \qquad I = \dfrac{4 \times 2^3}{12} = 2.67 \text{ in}^4$

$$= \frac{500 \times (16 \times 12)^3}{48 \times 30 \times 10^6 \times 2.67} + \frac{5 \times (100/12) \times (16 \times 12)^4}{384 \times 30 \times 10^6 \times 2.67} = 2.76"$$

4.33 a) $\delta = \theta L_2 = \dfrac{PL^2}{16EI} \times L_2. \quad \therefore 4 = \dfrac{P \times 240^2 \times 120}{16 \times 30 \times 10^6 \times \pi \times 1^4/4}. \quad \therefore P = 218 \text{ lb}$

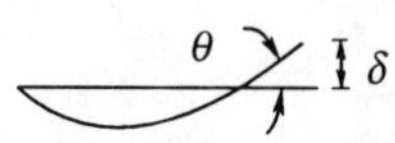

4.34 **d)** $\tau_{max} = \frac{1}{2}\sqrt{\left(\sigma_x - \sigma_y\right)^2 + 4\tau^2} = \frac{1}{2}\sqrt{6{,}000^2 + 4 \times 4{,}000^2} = 5{,}000$ psi

4.35 **a)** $\sigma_{max} = \frac{1}{2}\left(\sigma_x + \sigma_y\right) + \tau_{max} = 0 + 4{,}000 = 4{,}000$ psi

4.36 **c)** $\tau_{max} = \frac{1}{2}\sqrt{(3{,}000 + 5{,}000)^2 + 4 \times 3{,}000^2} = 5{,}000$ psi

4.37 **d)** $\tau = Tr/J = 500 \times 12 \times 1/\left(\pi \times 1^4/2\right) = 3{,}820$ psi.

$\sigma = P/A = 10{,}000/\pi \times 1^2 = 3{,}180$psi.

$\tau_{max} = \frac{1}{2}\sqrt{3{,}180^2 + 4 \times 3{,}820^2} = 4{,}137$ psi

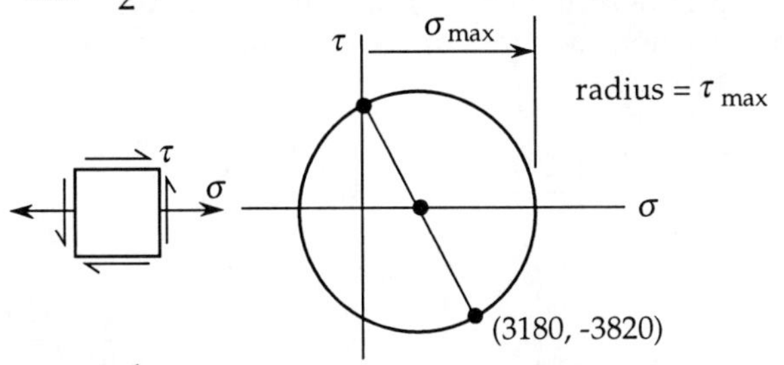

4.38 **c)** $\sigma_{max} = \frac{1}{2} \times 3{,}180 + 4{,}137 = 5{,}727$ psi

4.39 **b)** $M = 30{,}000$ in · lb. $\frac{My}{I} = \frac{30{,}000 \times 1}{\pi \times \left(1^4/4\right)} = 38{,}200$ comp.

$\frac{P}{A} = \frac{10{,}000}{\pi \times 1^2} = 3{,}183$ tens.

$\sigma_A = 38{,}200 - 3{,}183 = 35{,}000$ comp.

4.40 **c)** $\tau_{max} = \sigma/2 = 17{,}500$ psi. $VQ/Ib = 0$ on outer fibers

4.41 **c)** $\tau = \frac{Tr}{J} = \frac{2{,}000 \times 10 \times 1}{\pi \times \left(1^4/2\right)} = 12{,}732$ psi

$\sigma = \frac{My}{I} = \frac{2{,}000 \times 16 \times 1}{\pi \times \left(1^4/2\right)} = 40{,}744$ psi

$\tau_{max} = \frac{1}{2}\sqrt{40{,}744^2 + 4 \times 12{,}732^2} = 24{,}023$ psi

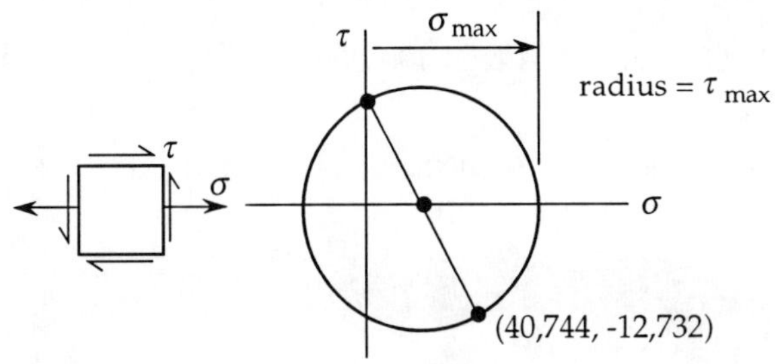

4.42 **d)** $\sigma_{max} = \frac{1}{2} \times 40{,}744 + 24{,}023 = 44{,}395$ psi

4.43 **c)** $M_{max} = FL_2 = 450 \times 3.28 = 1476$ ft · lb.

$$\sigma = \frac{My}{I} = \frac{(1476 \times 12) \times 1.97/2}{0.787 \times 1.97^3/12} = 34{,}790 \text{ psi.}$$

4.44 **d)** $\frac{VQ}{Ib} = \frac{450 \times (0.787 \times 0.985) \times 0.4925}{(0.787 \times 1.97^3/12) \times 0.787} = 435$ psi.

4.45 **a)** The force F provides a torque and a force acting on the end of the steel shaft. This is best described by (a) since the top fibers experience tension. There is no normal stress in the circumferential direction, ruling out (b), (c) and (d).

4.46 **e)** $\sigma = \frac{My}{I} = \frac{(450 \times 6.56 \times 12) \times 0.985}{\pi \times 0.985^4/4} = 47{,}200$ psi.

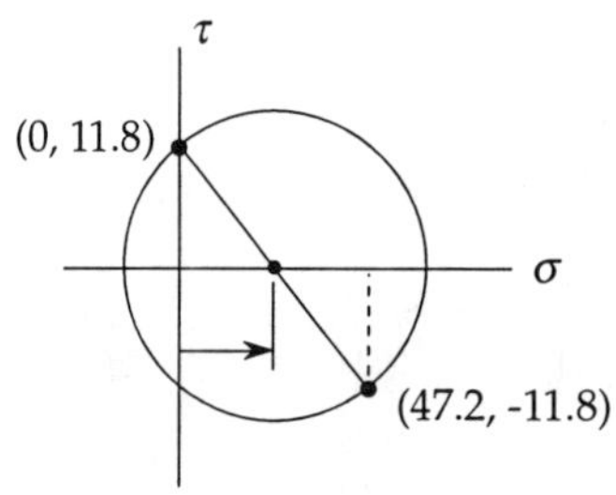

$$\tau = \frac{Tr}{J} = \frac{(450 \times 3.28 \times 12) \times 0.985}{\pi \times 0.985^4/2} = 11{,}800 \text{ psi.}$$

$$\therefore \sigma_{max} = 23{,}600 + \left[23{,}600^2 + 11{,}800^2\right]^{1/2}$$

$$= 50{,}000 \text{ psi.}$$

4.47 **d)** $\delta = \frac{PL^3}{3EI} = \frac{450 \times (6.25 \times 12)^3}{3 \times 30 \times 10^6 \times \pi \times 0.985^4/4} = 2.85$ in.

4.48 **b)** $\sigma_t = pD/2t. \quad \therefore p = 24{,}000 \times 2 \times \frac{1}{4}/24 = 500$ psi

4.49 **c)** $\sigma_a = pD/4t. \quad \therefore t = \frac{2000 \times 48}{4 \times 30{,}000} = 0.800$ in

4.50 **e)** $\tau_{max} = \frac{1}{2}\sqrt{(30{,}000 - 30{,}000)^2 + 4 \times 0^2} = 0$

4.51 **a)** $\left(\frac{\Delta L}{L}\right)_s = \left(\frac{\Delta L}{L}\right)_c. \quad \therefore \varepsilon_s = \varepsilon_c. \quad \therefore \sigma_s = \frac{E_s}{E_c}\sigma_c = 10\,\sigma_c.$

$F_s + F_c = 400{,}000$ or $A_s\sigma_s + A_c\sigma_c = 400{,}000.$

$$\sigma_s\left[\pi\left(5.5^2 - 5^2\right) + \pi \times \frac{5^2}{10}\right] = 400{,}000. \quad \therefore \sigma_s = 16{,}430 \text{ psi}$$

4.52 **d)** $n = E/E_{min} = 3$. The area is transformed:

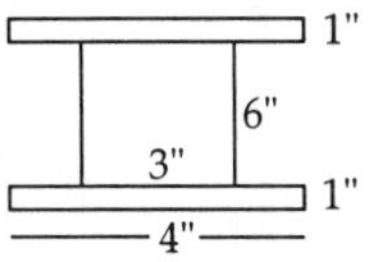

$$I_t = \frac{4 \times 8^3}{12} - \frac{1 \times 6^3}{12} = 152.7 \text{ in}^4.$$

$$\therefore \sigma_{al} = \frac{My}{I} = \frac{2,500 \times 12 \times 4}{152.7} = 786 \text{ psi}, \quad \sigma_s = \frac{nMy}{I} = \frac{3 \times 2,500 \times 12 \times 3}{152.7} = 1770 \text{ psi}$$

4.53 **d)** $n = 3$. The area is transformed:

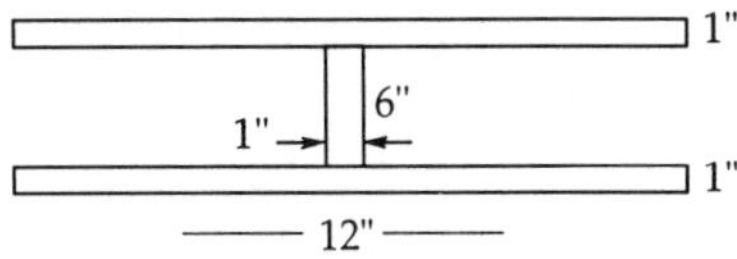

$$I_t = \frac{12 \times 8^3}{12} - \frac{11 \times 6^3}{12} = 314 \text{ in}^4.$$

$$\therefore \sigma_s = \frac{nMy}{I} = \frac{3 \times 2,500 \times 12 \times 4}{314} = 1,146 \text{ psi}.$$

4.54 **d)** $60 = \frac{L}{r} = \frac{L}{\sqrt{I/A}} = \frac{L}{\sqrt{4 \times \left(4^3/12\right)/16}}$. $\therefore L = 69.3"$ or $5.77'$

Assume $P = 1,000$ lb using a factor of safety of 2.

4.55 **c)** $1,000 = \frac{\pi^2 \times 10 \times 10^6 \times \pi \times 4^4/64}{4L^2}$. $\therefore L = 557"$ or $46.4'$

4.56 **e)** $P_{cr} = 4\pi^2 EI/L^2 = \alpha \Delta TEA$. $\Delta T = \frac{4\pi^2 I}{\alpha AL^2} = \frac{4\pi^2 \times \pi \times 1^4/64}{6.5 \times 10^{-6} \times \pi \times 0.5^2 \times 120^2} = 26.4\ °\text{F}$

4.57 **b)** $P_{cr} = 4\pi^2 EI/L^2 = 8,000$. $\therefore \pi^2 EI/L^2 = 2,000$.
$\therefore P_{cr} = \pi^2 EI/4L^2 = 2,000/4 = 500$ lb.

Thermodynamics

by Merle C. Potter

Solutions to Practice Problems

5.1 **b)** A tire. All other devices have fluid entering and/or leaving.

5.2 **d)** Mass. All other quantities do not depend on the mass.

5.3 **b)** Isobaric. The pressure will remain constant due to the inlets and outlets for air.

5.4 **a)** All properties are uniform throughout the volume.

5.5 **e)** A scientific law results from experimental observations.

5.6 **d)** Density does not depend on the mass, i.e., when the mass is increased the density does not increase simultaneously.

5.7 **e)** Distance, area, or volume may be measured in cm, cm^2, or cm^3, respectively, but grams are not acceptable in SI and cm cannot be used in combination with other units.

5.8 **a)** $F = Kx. \qquad P = \frac{AE}{L}\delta. \qquad \therefore K = \frac{AE}{L}.$

5.9 **e)** $\rho = 1/v = P/RT = (-40+100)/.287 \times (-40+273) = 0.897 \ \text{kg/m}^3.$
$R = 8.314/28.97 = .287.$

5.10 **b)** $v = v_f + x(v_g - v_f)$ $P = 0.2$ MPa

$0.5 = .001 + x(.8857 - .001)$ TABLE 5.3.2

$\therefore x = 0.564$

5.11 **d)** If the state is beyond Table 5.3.3 use the equation of state with

$R = \frac{8.314}{18} = 0.462.$

$$v = \frac{RT}{P} = \frac{0.462 \times 1473}{4000} = 0.17 \text{ m}^3/\text{kg}.$$

Use $T = 1200 + 273 = 1473$ K, $P = 4000$ kPa

5.12 **a)** In a rigid vessel, the volume remains constant. If we move straight up on the *T-v* diagram, *x* increases. (Temperature increases as heat is added.) Note: it is important to know that the diagram is skewed quite markedly to the left.

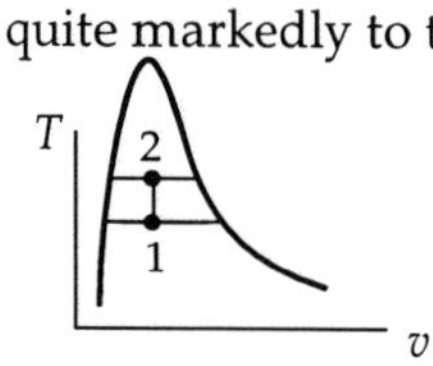

5.13 **b)** $v = \frac{V}{m} = \frac{2}{20} = 0.1,$ $P = 4$ MPa.

From Table 5.3.3 interpolation gives

$T = \frac{0.1 - 0.09885}{0.1109 - 0.09885} \times 100 + 600 = 610°$ C.

5.14 **a)** Using Table 5.3.2 we see that 115°C < 120.2°C. $\therefore$ compressed.

5.15 **c)** $m = \frac{P_1V_1}{RT_1} = \frac{280 \times .03}{.287 \times 263} = 0.1113$ kg, where $R = \frac{8.3.4}{28.97} = 0.287.$

$\therefore V_2 = \frac{mRT_2}{P_2} = \frac{.1113 \times .287 \times 303}{310} = 0.0312 \text{ m}^3$

5.16 **e)** $\dot{Q} = \frac{1}{R}A\Delta T$ $A = 4\pi r^2$

$\frac{300}{1000} = \frac{1}{1.5}4\pi \times 5^2(T - 20)\frac{1}{3600}.$ $\therefore T = 25.2°\text{C}.$

The 1000 converts J/s to kJ/s.

The 3600 converts hr to seconds.

5.17 **e)** $W = F \times d = (10 \times 9.8) \times 3 = 294$ N·m input. $\therefore$ W = -294 N·m.

$\therefore Q = W = -294$ J. (Heat output is negative.)

5.18 **d)** $Q - W = E_a - E_b.$

$20 - 5 = E_a - 10.$ $\therefore E_a = 25$ kJ.

5.19 **a)** For $2 \rightarrow 3$ $Q-(-5)=5.$ $\therefore Q_{2-3}=0$

$\therefore\ Q_{net}=20+0+30=50$

$Q_{net}=W_{net}=5-5+W_{3-1}=50.$

$\therefore\ W_{3-1}=50\ \text{kJ}.$

5.20 **c)** $Q=mc\Delta T$. Here $Q=\frac{1}{2}mV^2,\quad m=V\rho$

$\frac{1}{2}\times 2000\times 25^2=10\ 000\times 10^{-6}\times 1000\times 4180\ \ \Delta T$

$\therefore\ \Delta T=14.95°\text{C}$

5.21 **b)** $(mc\Delta T)_{ice}+(mc\Delta T)_{\substack{melted\\ice}}+mh_{fusion}=(mc\Delta T)_{water}$

$10\times 2.1\times 10+10\times 4.18(T-0)+10\times 320=100\times 4.18(20-T).\quad \therefore T=10.77°\text{C}.$

5.22 **b)** The heat transfer exceeds the work done if $\Delta E>0$.

5.23 **b)** δW and δQ are inexact differentials. That's why we don't write dW and dQ.

5.24 **a)** $W=mRT\ln P_1/P_2$

$=2\times .287\times 573\ln 100/4000=-1213\ \text{kJ}$

5.25 **b)** $\Delta U=m(u_2-u_1)$ Table 7.3.3

$=2(2725.3-2810.4)=-170.2\ \text{kJ}$

5.26 **e)** $Q=m\Delta h=mh_{fg}=2\times 1940.8=3881.6\ \text{kJ}$

5.27 **e)** Assume P = const since doors would be opened. Cracks abound.

$$Q=mC_p\Delta T\ .$$

$$\frac{400\times 200}{3600}+\frac{30\times 2000}{1000}=\frac{6000\times 1.23\times 1.00\Delta T}{20\times 60}.\qquad \therefore\ \Delta T=13.37°\ \text{C}$$

5.28 **d)** $Q=m\Delta h=mc_p\Delta T$

$522=1\times c_p\times(800-300).\quad \therefore\ c_p=1.044\frac{\text{kJ}}{\text{kg}\cdot\text{K}}$

5.29 **e)** $Q=m\Delta h=mc_p\Delta T=10\times 1.00\times(230-10)=2200\ \text{kJ}$

5.30 **c)** $Q=m\Delta h=10[3072-42]=30\,300\ \text{kJ}$

5.31 **a)** $W=P\Delta V=mP\Delta v=10(1.316-.001)\times 200=2630\ \text{kJ}$

5.32 **a)** $Q=m\Delta u$ $\qquad u_1=851+.5(2595-851)=1723\ \text{kJ/kg}.$

$T_2=500°\text{C}$

$v_2=.0642$ $\qquad v_1=.001+.5(.1274-.001)=.0642\ \text{m}^3/\text{kg}.$

Use Table 5.3.3

$u_2=\frac{.0642-.0566}{.0686-.0566}(3091-3082)+3082=3088$

$\therefore Q=2(3088-1723)=2730\ \text{kJ}$

5.33 **d)** $\frac{P_1}{T_1}=\frac{P_2}{T_2}$. $(p_2+100)273=(100+100)343$

$\therefore P_2 = 151$ kPa gage

5.34 **e)** $Q = m\Delta u = mc_v\Delta T$ $m=\frac{PV}{RT}=\frac{100\times 10}{.287\times 293}=11.89$ kg

$= 11.89\times .716\times 60 = 511$ kJ

5.35 **e)** $Q = m\Delta u$. $9000 = 10(u_2 - 42)$ $\therefore u_2 = 942$ kJ/kg

$942 = 851 + x_2(2595 - 851)$ $\therefore x_2 = .0522$

5.36 **d)** $T_2 = T_1\left(\frac{P_2}{P_1}\right)^{k-1/k} = 473\left(\frac{20}{400}\right)^{0.4/1.4} = 201\text{K}$ or $-72°\text{C}$

5.37 **a)** $T_2 = T_1\left(\frac{v_1}{v_2}\right)^{k-1} = 473\left(\frac{1}{3}\right)^{0.4} = 305\text{K}$ or $31.8°\text{C}$

5.38 **c)** $s_2 = s_1 = 7.1677$ from Table 5.3.2, $7.1677 = .649 + 7.502x_2$. $\therefore$ $x_2 = .869$

5.39 **b)** $s_2 = s_1 = 7.1685$ from Table 5.3.3. This is slightly less than $s = 7.171$ at $T = 200°\text{C}$ and $P = 0.4\,\text{MPa}$. $\therefore T_2 = 199°\text{C}$.

5.40 **b)** $T_2 = T_1\left(\frac{P_2}{P_1}\right)^{k-1/k} = 293\left(\frac{6000}{100}\right)^{.4/1.4} = 944$ K.

$W = -m\Delta u = -mc_v\Delta T$ (let $m = 1$)

$= -.716(944 - 293) = -466$ kJ/kg .

5.41 **d)** The heat transfer must be zero.

5.42 **e)** $T_2 = T_1\left(\frac{P_1}{P_2}\right)^{n-1/n} = 283\times 3^{.2/1.2} = 340$ K or $66.9°\text{C}$

5.43 **a)** $\rho_1 = \frac{1}{v_1} = \frac{1}{.06525} = 15.33$ Table 7.3.3

$\rho_2 = \frac{1}{v_2} = \frac{1}{7.649} = 0.1307$ Table 7.3.2

$\rho_1 A_1 V_1 = \rho_2 A_2 V_2$. $15.33\pi\times\frac{.2^2}{4}V_1 = .1307\pi\times\frac{.05^2}{4}V_2$. $\therefore$ $\frac{V_2}{V_1} = 1877$

5.44 **e)** $Q = h_2 - h_1$. $h_1 = 251$ from Table 5.3.1.

$= 3674 - 251 = 3423$ kJ/kg

5.45 **b)** $\dot{Q} = \dot{m}_w c\Delta T = \dot{m}_s \Delta h$. Table 5.3.1, $h_{fg} = 2358$ kJ/kg.

$\dot{m}\times 4.18(30-20) = 10\times 2358$. $\therefore \dot{m} = 564$ kg/s

5.46 **c)** $s_2 = s_1 = 6.7698 = .649 + 7.502x_2$. $\therefore x_2 = .816$.

$\therefore h_2 = 192 + .816\times 2393 = 2144$. $W = h_1 - h_2 = 3213.5 - 2144 = 1069$ kJ/kg

5.47 **e)** $\dot{W}_P = \dot{m}\dfrac{\Delta P}{\rho\eta} = 10\dfrac{6000-10}{1000\times.75} = 79.9$ kW

5.48 **e)** $\dot{W}_T = \dot{m}\dfrac{\Delta P}{\rho} = 60\times2\times2\times1000\dfrac{300}{1000} = 72\ 000$ kW

5.49 **e)** $T_2 = T_1\left(\dfrac{P_2}{P_1}\right)^{k-1/k} = 673\left(\dfrac{80}{2000}\right)^{.4/1.4} = 268$ K.

$V_2^2/2 = h_1 - h_2 = c_p(T_1 - T_2) = 1000(673-268)$. $\therefore V_2 = 900$ m/s.

5.50 **b)** $h_1 = 3177$ $s_{2'} = s_1 = 6.5415 = .832 + x_{2'}7.0774$. $\therefore x_{2'} = .807$.

$\therefore h_{2'} = 251 + .807\times2358 = 2153$. $w = 0.87\times(3177-2153) = 891$ kJ/kg.

5.51 **b)** $\eta_{\max} = 1 - \dfrac{T_L}{T_H} = 1 - \dfrac{293}{393} = 0.254$

$\eta = \dfrac{4180}{10\times4.18(120-20)} = 1.00$. $\therefore$ impossible

5.52 **b)** $\Delta s_{net} > 0$ for all processes.

5.53 **d)** Operates between const. temp. reservoirs.

5.54 **d)** $\Delta S > Q/T = 90\times320/273 = 105.5$ kJ/K

5.55 **e)** $40(320 + 4.18T_2) = 100\times4.18(20 - T_2)$. $\therefore T_2 = -7.6°$C.

Impossible. $\therefore T_2 = 0°$C. $320\ m = 100\times4.18\times20$. $\therefore m = 26.1$ kg.

$\Delta s = 26.1\times320/273 + 100\times4.18\ln 273/293 = 1.04$

5.56 **b)** $\eta_{\max} = 1 - \dfrac{T_L}{T_H} = 1 - 323/873 = 0.63$

5.57 **e)** $T_4 = P_4 v_4/R = 160\times.5/.287 = 279$ K.

$w = q\eta = 30(1 - 279/473) = 12.3$ kJ/kg

5.58 **a)** $\eta = 1 - (400/40)^{-.4} = 0.602$. $\therefore \dot{Q} = \dot{W}/\eta = 100/.602 = 166$ kW.

5.59 **a)** $r_C = 45/25 = 1.8$. $r = 450/25 = 18$.

$\dot{Q} = \dot{W}/\eta = 120\Big/\left[1 - \dfrac{1.8^{1.4}-1}{1.4(1.8-1)}18^{-.4}\right] = 187$ kJ/s.

5.60 **c)** $s_4 = s_3 = 7.1677 = .649 + 7.502x_4$. $\therefore x_4 = .869$.

$h_4 = 192 + .869\times2393 = 2272$. $h_3 = 3658$, $h_1 = 192$.

$\eta = \dfrac{w}{q} = (3658-2272)/(3658-192) = 0.40$

5.61 **c)** $s_4 = s_3 = 7.168$. $\therefore$ ④ superheat $\therefore h_4 = 2722$. $h_3 = 3658$.

After reheat $h_5 = 3704$. $s_6 = s_5 = 8.777$.

$$\therefore h_6 = \frac{8.777 - 8.448}{8.9038 - 8.448}(2879.5 - 2688) + 2688 = 2826.$$

$$\eta = \frac{w_{3-4} + w_{5-6}}{q_{2-3} + q_{4-5}} = \frac{3658 - 2722 + 3704 - 2826}{3658 - 192 + 3704 - 2722} = 0.408$$

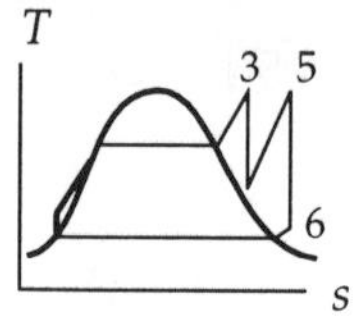

5.62 **e)** Decrease moisture content in the turbine.

5.63 **c)** $s_2 = s_1 = 1.75$. Refer to Fig. 5.4a: $h_1 = 380$, $h_2 = 430$, $h_3 = 255 = h_4$.

$$COP = \frac{q_{4-1}}{w_{1-2}} = \frac{380 - 255}{430 - 380} = 2.5.$$

5.64 **e)** $h_4 = h_3 = 255 = h_f + x_4(h_g - h_f)$

$$= 165 + x_4(380 - 165). \quad \therefore x_4 = 0.42.$$

5.65 **a)** $\dot{W}_{max} = \dot{m}(h_1 - T_o s_1) - \dot{m}(h_2 - T_o s_2)$

$$= \dot{m}[(h_1 - h_2) + T_o(s_2 - s_1)] = \dot{m}\left[c_p \Delta T + T_o\left(c_p \ln\frac{T_2}{T_1} - R\ln\frac{P_2}{P_1}\right)\right]$$

$$= 10\left[1.00(100 - 20) + 293\left(1.00\ln\frac{293}{373} - .287\ln\frac{80}{6000}\right)\right] = 3723 \text{ kW}$$

5.66 **d)** $\phi_2 - \phi_1 = w_{max}$.

5.67 **b)**

5.68 **b)**

5.69 **c)** $\dot{Q} = .13 \times 2 \times 8 \times \frac{25}{.5} = 104$ J/s

5.70 **a)** $\frac{1\,500\,000}{10 \times 60} = h \times \pi \times .04 \times (150 - 40)$. $\therefore h = 181$ J/s·m^2·°C

5.71 **d)** $100 = 5.67 \times 10^{-8} \times (.8 \times 10^{-4})(T^4 - 353^4)$.

$\therefore T = 2167$ K or 1894° C

5.72 **b)** $R = \frac{1}{h} + \frac{L_1}{k_1} + \frac{L_2}{k_2} + \frac{L_3}{k_3} + \frac{1}{h}$

$$= \frac{1}{10} + \frac{.025}{.04} + \frac{.08}{.038} + \frac{.02}{.15} + \frac{1}{10} = 3.06 \ \frac{\text{s}\cdot\text{m}^2\cdot°\text{C}}{\text{J}}$$

5.73 **e)** $\omega = 0.0118$. $\therefore m_v = \omega m_a = \omega V/v$

$$= .0118 \times 600/.861$$

$$= 8.2 \text{ kg} \quad \text{or } 8.2 \text{ L}$$

5.74 **d)** The enthalpy remains constant since $\dot{W} = \dot{Q} = 0$. States 1 and 2 lie on a constant enthalpy line on the psychrometric chart with $\phi_2 = 100\%$. $\therefore T_2 = 19°$C.

Fluid Mechanics

by Merle C. Potter

Solutions to Practice Problems

6.1 **b)** a) is true for a liquid and low speed gas flows. c) and e) are true of gases. d) is true for a liquid.

6.2 **d)** $\tau=\mu\, du/dy.\quad \therefore\ \mu=\tau/du/dy.\quad [\mu]=\dfrac{F/L^2}{\frac{L}{T}/L}=\dfrac{FT}{L^2}=\dfrac{\left(ML/T^2\right)T}{L^2}=\dfrac{M}{LT}$

6.3 **a)** Viscosity μ varies with temperature only.

6.4 **e)** $dp=-\gamma dz=-\rho g dz.\quad p=\rho RT$ (ideal gas)

$$\therefore\ dp=-\frac{p}{RT}gdz \quad \text{or} \quad \frac{dp}{p}=-\frac{g}{RT}dz.$$

$$\int\frac{dp}{p}=-\frac{g}{RT}\int dz.\quad \therefore\ \ln p=-Cz.\quad \therefore\ p=e^{-Cz}.$$

6.5 **e)**

6.6 **a)** $T=\tau Ar=\mu\dfrac{du}{dy}Ar=\mu\dfrac{r\omega}{t}2\pi rLr.$

$$1.6=\mu\frac{0.04\times1000}{0.001}2\pi\times0.04\times0.04\times0.04.\quad \therefore\ \mu=0.1\ \text{N}\cdot\text{s/m}^2$$

6.7 **c)** $K=-\mathcal{V}\dfrac{\Delta p}{\Delta\mathcal{V}}=-2\dfrac{500}{-0.004}=250\ 000$ or 250 MPa.

6.8 **c)** $\sigma 2\pi D = \gamma\pi r^2 L$

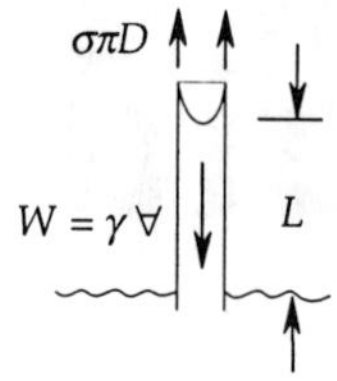

$0.0736 \times \pi \times 0.001 = 9800\pi \times 0.0005^2 L$

$\therefore L = 0.03$ m or 3 cm.

6.9 **e)** Cavitation occurs when the pressure reaches the vapor pressure = 2.45 kPa. (See Table 6.2.)

6.10 **c)** $L = V\Delta t = \sqrt{kRT}\Delta t = \sqrt{1.4 \times 287 \times 288} \times 1.2 = 408$ m.

Assume $T = 15°C$. T must be in absolute. $\therefore T = 288$.

6.11 **a)** $v = \frac{\mu}{\rho} = \frac{0.0034}{1.3 \times 1000} = 2.6 \times 10^{-6}$.

6.12 **a)** $p = \gamma_1 \Delta h_1 + \gamma_2 \Delta h_2$ where $\gamma_2 = SG \times \gamma_{H_2O}$

$= 9800 \times 2 + 1.04 \times 9800 \times 4 = 60\ 400$ Pa or 60.4 kPa.

6.13 **c)** $p = \gamma h$ where $\gamma_{Hg} = 13.6\ \gamma_{H_2O}$

$= 13.6 \times 9800 \times 0.6 = 80\ 000$ Pa.

6.14 **d)** $p = \gamma h = 1.03 \times 9800 \times 1200 = 12.1 \times 10^6$ Pa or 12.1 MPa.

6.15 **c)** $dp = -\gamma dz = -\rho g dz = -p\frac{g}{RT}dz$ using $p = \rho RT$

$\therefore \frac{dp}{p} = -\frac{g}{RT}dz.\ \int_{100}^{p} dp/p = -\frac{g}{RT}\int_0^{2000} dz.$

$\therefore \ln\frac{p}{100} = -\frac{9.8}{287 \times 293} \times 2000.\ \therefore p = 79.2$ kPa.

6.16 **c)** $F = pA.\ p = \frac{1000}{\pi \times 0.05^2} + 0.86 \times 9800 \times 0.2 = 129\ 000$ Pa.

$\therefore F = 129\ 000 \times \pi \times 0.0025^2 = 2.53$ N.

6.17 **a)** $p = \gamma h = (13.6 \times 9800) \times 0.2 = 26\ 700$ Pa or 26.7 kPa.

6.18 **e)** $p + 9800 \times 0.3 = 13.6 \times 9800 \times 0.3$. $\therefore p = 37\ 000$ Pa.

6.19 **c)** $F = pA = (20\ 000 + 9800 \times 3)\pi \times 1^2 = 155\ 000$ N.

6.20 **b)** $F = p_c A = 9800 \times \frac{1.5}{2} \times 1.5^2 = 16\ 500$ N.

6.21 **d)** $7P = \frac{5}{3}F = \frac{5}{3}\gamma h_c A$. $\therefore P = \frac{5}{21} \times 9800 \times 2 \times 15 = 70\ 000$ N.

6.22 **d)** All pressures on the curved section pass through the center. Moments about the hinge give

$P = F_v = \gamma \times \text{Volume} = 9800 \times (9\pi/4) \times 4 + 9800 \times 6 \times 3 \times 4 = 983\ 000$ N.

6.23 **b)** $y_p = y_c + \frac{I_c}{y_c} = 6 + \frac{5\times 4^3/12}{6\times 20} = 6.22\text{ m}.$

$4P = F\times 2.22 = \gamma h_c A\times 2.22. \quad \therefore P = 9800\times 6\times 20\times 2.22/4 = 653\ 000\text{ N}.$

6.24 **e)** $W = \gamma\forall. \quad 4\times 1500\times 9.8 = 9800\times 6\times 12\times h. \quad \therefore h = 0.0833\text{ m}.$

6.25 **c)** $25 = 100 - 9800\forall. \quad \therefore \forall = 0.00765\text{ m}^3.$

$100 = 9800(SG)\times 0.00765. \quad \therefore SG = 1.33.$

6.26 **d)** $\Delta p = \Delta\gamma\times h = \left(\frac{1}{253} - \frac{1}{293}\right)\times\frac{100}{0.287}\times 3\times 9.8 = 5.53\text{ Pa}.$

6.27 **d)** $[p] = \frac{M}{LT^2} \quad [Q] = \frac{L^3}{T} \quad [D] = L \quad [\rho] = \frac{M}{L^3}.$

First, eliminate M, then T, then L:

$$\frac{M}{LT^2}\cdot\frac{L^3}{M}\cdot\frac{T^2}{L^6}\cdot L^4 = p\cdot\frac{1}{\rho}\cdot\frac{1}{Q^2}\cdot D^4 = \frac{pD^4}{\rho Q^2}.$$

6.28 **a)** $[\sigma] = \frac{M}{T^2} \quad [\rho] = \frac{M}{L^3} \quad [D] = L \quad [V] = \frac{L}{T}.$ Combine: $\frac{\sigma}{\rho}\frac{1}{V^2}\frac{1}{D}$

6.29 **b)** Inertial force to viscous forces.

6.30 **d)**

6.31 **d)** $(\text{Fr})_m = (\text{Fr})_p. \quad \therefore \frac{V_m^2}{l_m g} = \frac{V_p^2}{l_p g}. \quad \therefore \frac{V_m^2}{V_p^2} = \frac{1}{20}.$

$Q_m^* = Q_p^*$ or $\frac{Q_m}{V_m l_m^2} = \frac{Q_p}{V_p l_p^2}. \quad \therefore Q_m = 4\times\frac{1}{20^2}\times\frac{1}{\sqrt{20}} = 0.0022.$

6.32 **a)** $\text{Re}_m = \text{Re}_p. \quad \left(\frac{Vl}{\nu}\right)_m = \left(\frac{Vl}{\nu}\right)_p. \quad \therefore \frac{V_m}{V_p} = \frac{l_p}{l_m} = 10.$

$\therefore V_m = 10\,V_p = 10\times 10 = 100\text{ m/s}.$

6.33 **a)** $\text{Fr}_m = \text{Fr}_p. \quad \therefore \left(\frac{V^2}{lg}\right)_m = \left(\frac{V^2}{lg}\right)_p. \quad \therefore \frac{V_p^2}{V_m^2} = \frac{\ell_p}{\ell_m}.$

$(F_D)_m^* = (F_D)_p^*$ or $\frac{(F_D)_m}{\rho_m V_m^2 l_m^2} = \frac{(F_D)_p}{\rho_p V_p^2 l_p^2}. \quad \therefore (F_D)_p = 10\frac{\rho_p}{\rho_m}\frac{V_p^2}{V_m^2}\frac{l_p^2}{l_m^2} = 10\frac{l_p^3}{l_m^3}.$

$\therefore (F_D)_p = 10\times 40^3 = 640\ 000\text{ N}.$

6.34 **d)** $\mathrm{Fr}_m = \mathrm{Fr}_p. \quad \left(\frac{V^2}{lg}\right)_m = \left(\frac{V^2}{lg}\right)_p. \quad \therefore \frac{V_p^2}{V_m^2} = 10.$

$$\dot{W}_p^* = \dot{W}_m^*. \qquad \frac{\dot{W}_m}{\rho_m V_m^3 l_m^2} = \frac{\dot{W}_p}{\rho_p V_p^3 l_p^2}.$$

$$\therefore \dot{W}_p = \frac{V_p^3}{V_m^3}\frac{l_p^2}{l_m^2}\dot{W}_m = 10^3\sqrt{10} \times 20 = 63\ 250\ \mathrm{W}.$$

6.35 **e)** $V_2 = 20\pi \times 2^2 / \pi \times 5^2 = 3.2\ \mathrm{m/s}.$

6.36 **d)** $V_2 = 20\pi \times 1^2 / 100 \times \pi \times 0.1^2 = 20\ \mathrm{m/s}.$

6.37 **e)** $V_2 \times 2\pi \times 40 \times 0.2 = 20 \times \pi \times 1^2. \quad \therefore V_2 = 1.25\ \mathrm{m/s}.$

6.38 **a)** Bernoulli's equation is only applicable in a steady, viscous, incompressible, flow along a streamline in an inertial coordinate system. It must be modified to accommodate other effects, such as compressible viscous flow.

6.39 **b)** $p = \rho V^2/2 = 1.23 \times 25^2/2 = 384\ \mathrm{Pa}. \quad F = pA = 384 \times \pi \times 0.075^2 = 6.79\ \mathrm{N}.$

6.40 **e)** $V_2 A_2 = V_1 A_1. \quad \therefore V_2 = V_1 \times 4^2/2^2 = 4V_1.$

$$\frac{p_1}{\rho} + \frac{V_1^2}{2} = \frac{p_2}{\rho}^{\nearrow 0} + \frac{V_2^2}{2}. \qquad \frac{700\ 000}{1000} + \frac{V_1^2}{2} = \frac{16V_1^2}{2}. \qquad \therefore V_1 = 9.66\ \mathrm{m/s}$$

6.41 **d)** $p + \rho\frac{V^2}{2} + 9800 \times 0.1 = p + 13.6 \times 9800 \times 0.1.$

$$\therefore V^2 = 12.6 \times 9800 \times 0.1 \times 2/1000. \quad \therefore V = 4.97\ \mathrm{m/s}.$$

6.42 **c)** Cavitation results if $p_2 = -100\ \mathrm{kPa}.$

$$\frac{p_1}{\rho} + \frac{V_1^2}{2}^{\nearrow 0} = \frac{p_2}{\rho} + \frac{V_2^2}{2}. \quad 900\ 000/1000 = -100\ 000/1000 + V_2^2/2. \quad \therefore V_2 = 44.7\ \mathrm{m/s}.$$

6.43 **a)** $F = \rho A V^2 = 1.2 \times \pi \times 0.01^2 \times 80^2 = 2.41.$

6.44 **b)** Momentum: $-F = \rho A v_r (v_{r2} - v_{r1})_x. \quad F = 1000 \times 0.02 \times 10(10 - 0) = 2000\ \mathrm{N}.$

(v_r is relative speed)

Note: With a single blade we must also use relative speed for $\dot{m} = \rho AV$ since not all of the fluid leaving the jet has its momentum changed.

6.45 **c)** $F = \rho A V^2 = .5 \times \pi \times 0.25^2 \times 1200^2 = 141\ 000\ \mathrm{N}.$

6.46 **a)** $-F = \rho A V(-V - V). \quad \therefore F = 2\rho A V^2.$

$$\therefore F = 2 \times 1000 \times 0.05 \times 0.8 \times 50^2 = 200\ 000\ \mathrm{N}.$$

6.47 **c)** $p_1A_1 - F = \rho A_1 V_1(V_2 - V_1)$. $V_2 = 4V_1$. $\frac{p_1}{\rho} + \frac{V_1^2}{2} = \frac{p_2}{\rho} + \frac{V_2^2}{2} = \frac{16V_1^2}{2}$. (with $p_2 = 0$)

$\therefore V_1 = \sqrt{\frac{800\ 000}{1000} \times \frac{2}{15}} = 10.3.$ $V_2 = 41.2$.

$\therefore F = 800\ 000 \times \pi \times 0.04^2 - 1000 \times \pi \times 0.04^2 \times 10.3 \times (41.2 - 10.3) = 2420\text{ N}.$

6.48 **b)** The energy grade line.

6.49 **a)** EGL has elevation $\frac{V^2}{2g} + \frac{p}{\gamma} + z$. The HGL has elevation $\frac{p}{\gamma} + z$.

6.50 **a)** $\dot{W}_P = \gamma Q \frac{\Delta p}{\gamma} \Big/ 0.85 = 0.2 \times 500 / 0.85 = 117.6\text{ kW}.$

6.51 **e)** $\dot{W}_T = \gamma Q \frac{\Delta p}{\gamma} \times 0.85 = 0.8 \times 600 \times 0.85 = 408\text{ kW}.$

6.52 **a)** manometer: $p_1 = p_2 + \rho V_2^2/2$. $-\frac{\dot{W}_T}{\gamma Q} = \frac{V_2^2}{2g} + \frac{p_2}{\gamma} - \frac{V_1^2}{2g} - \frac{p_1}{\gamma}$ (100% efficient)

$\therefore \dot{W}_T = Q\frac{V_2^2}{2}\rho\eta = (20 \times \pi \times 0.1^2)\frac{20^2}{2} \times 1000 \times .88 = 111\ 000\text{ W}.$

6.53 **c)** In an orifice, the minimum pressure (maximum velocity) occurs at the vena contracta. The loss coefficient is based on this minimum pressure.

6.54 **d)** The energy equation for a control volume, with no work or heat transfer demands that $\Delta h = 0$.

6.55 **c)** Laminar flow can occur in a pipe upwards of 40 000 Reynolds number. At low Reynolds number a laminar flow occurs in even a rough pipe. Viscous effects dominate in a laminar flow.

6.56 **b)** Increases with the velocity squared.

6.57 **d)** Increases linearly to the wall.

6.58 **c)** Vary as the 1/7th power law.

6.59 **a)** The flow is laminar and steady, and $v = V$ at $y = \delta$ and $v = 0$ at $y = 0$.

6.60 **e)** By curve E.

6.61 **b)** By curve B.

6.62 **a)** Pressure varies linearly. $\therefore \frac{\Delta p}{\Delta x} = \text{Const.}$

6.63 **e)** The Darcy-Weisbach equation.

6.64 **d)** Found by using loss coefficients.

6.65 **d)** Shear stress varies linearly with radius.

6.66 **b)** $h_f = f\frac{L}{D}\frac{V^2}{2g}$. $\therefore f = 40\frac{0.1}{100}\frac{2\times 9.8}{6^2} = 0.0218$.

6.67 **e)** $V = Q/A = \frac{0.02}{\pi\times 0.03^2} = 7.07 \text{ m/s}$. $\text{Re} = \frac{VD}{\nu} = 7.07\times 0.06/10^{-6} = 4.2\times 10^5$.

$\frac{e}{D} = \frac{.26}{60} = .0043$. From Fig. 11.2 $f = 0.03$.

$$\dot{W}_p = \frac{\gamma Q}{\eta}\left(\frac{V_2^2}{2g}^{\,0} + \frac{p_2}{\gamma}^{\,0} + z_2 - \frac{V_1^2}{2g}^{\,0} - \frac{p_1}{\gamma}^{\,0} - z_1 + f\frac{L}{D}\frac{V^2}{2g} + C\frac{V^2}{2g}\right)$$

$$= \frac{9800\times 0.02}{0.85}\left[80 - 20 + \left(0.03\frac{100}{.06} + 1 + .5\right)\frac{7.07^2}{2\times 9.8}\right] = 44\ 000 \text{ W}.$$

6.68 **e)** $V = Q/A = \frac{0.006}{\pi\times 0.03^2} = 4.77 \text{ m/s}$. $\text{Re} = \frac{VD}{\nu} = 4.77\times .04/10^{-6} = 1.9\times 10^5$.

$\frac{e}{D} = \frac{0.046}{40} = 0.0011$. From Fig. 4.2 $f = 0.022$. $0 = \frac{p_B - p_A}{\gamma} + f\frac{L}{D}\frac{V^2}{2g} + C\frac{V^2}{2g}$.

$\therefore p_B = 510\ 000 - \left(.022\frac{50}{.04} + 6.4 + 2\times .9\right)\frac{4.77^2}{2\times 9.8}\times 9800 = 104\ 000 \text{ kPa}$.

6.69 **e)** $V = Q/A = 0.3/(.15\times .4) = 5 \text{ m/s}$. $R_H = \frac{40\times 15}{110} = 5.45 \text{ cm}$.

$\text{Re} = \frac{5\times 4\times .0545}{1.6\times 10^{-5}} = 6.8\times 10^4$. With $\frac{e}{D} = 0$, $f = 0.02$.

$\Delta p = f\frac{L}{4R_H}\frac{V^2}{2g}\gamma = .02\frac{500}{4\times .0545}\frac{5^2}{2}\times 1.23 = 705 \text{ Pa}$.

6.70 **d)** $A_c/A_2 = .62 + .38(.5)^3 = 0.668$. $C_1 = (1 - .668)^2 = 0.11$.

$0.11\, V_c^2/2g = C\, V_2^2/2g$. $\therefore C = 0.11\left(\frac{A_2}{A_c}\right)^2 = .11\times\frac{1}{.668^2} = 0.25$.

6.71 **e)** $V = Q/A = 0.02/\pi\times .03^2 = 7.07 \text{ m/s}$. $e/D = .15/60 = .0025$

$\text{Re} = \frac{7.07\times .06}{10^{-6}} = 4.2\times 10^5$. $\therefore f = 0.024$.

$L_e = CD/f = 0.9\times 0.06/0.025 = 2.16 \text{ m}$.

6.72 **a)** $Q = \frac{1}{n}AR_H^{2/3}S^{1/2} = \frac{1}{.012}\times 6\times .86^{2/3}\times .001^{1/2} = 14.3$

where $R_H = 6/(3+4) = 0.86 \text{ m}$.

6.73 **b)** $Q = \frac{1}{n}AR_H^{2/3}S^{1/2} = \frac{1}{.016}4h\left(\frac{4h}{4+2h}\right)^{2/3}\times .001^{1/2} = 10$.

Trial-and-error: $h = 1.4 \text{ m}$.

6.74 **b)** $Q=\frac{1}{n}AR_H^{2/3}S^{1/2}=\frac{1}{.016}\pi\times1^2\times.5^{2/3}S^{1/2}=10,$ where $R_H=\frac{A}{P}=\frac{\pi\times1^2}{2\pi}=.5.$

$\therefore S=0.00654.$

6.75 **e)** $\rho=p/RT=500/.287\times313=5.57\ \text{kg/m}^3$

$\dot{m}=\rho AV=5.57\times\pi\times.05^2\times100=4.37\ \text{kg/s}$

6.76 **e)** $T_e=T_o(p_e/p_o)^{k-1/k}=293\left(\frac{100}{500}\right)^{.286}=185\,\text{K}.$

$V=\text{M}c=1\sqrt{1.4\times287\times185}=273\ \text{m/s}.$

6.77 **a)** $T_e=T_o(p_e/p_o)^{k-1/k}=293\left(\frac{100}{800}\right)^{.286}=162\,\text{K}.\ \ \therefore T_e=162-273=-111°\text{C}.$

6.78 **c)** To obtain supersonic flow from a reservoir the subsonic velocity must increase in a converging section until $m = 1$ at the throat and then continue to increase in the diverging section. As velocity increases temperature (and enthalpy) decrease. Velocity decreases in a converging section if $M > 1$.

6.79 **a)** $\text{M}_1=\frac{V_1}{c_1}=\frac{700}{\sqrt{1.4\times287\times303}}=2.01\quad\therefore\frac{p_2}{p_1}=4.54$ (from Normal Shock Table)

$\therefore p_1=p_2/4.54=500/4.54=110\ \text{kPa}$

6.80 **e)** $\frac{A}{A^*}=\frac{10^2}{6^2}=2.78.\quad\therefore 2.5<\text{M}_1<2.6$ (Isentropic flow Table)

6.81 **b)** $\sin\phi=\frac{1}{\text{M}}=\frac{1}{2}.\quad\therefore\phi=30°$

$\tan30°=\frac{1000}{L}.\quad\therefore L=1732\ \text{m}$

$\Delta t=\frac{L}{V}=\frac{1732}{2\sqrt{1.4\times287\times293}}=2.52\ \text{sec}.$

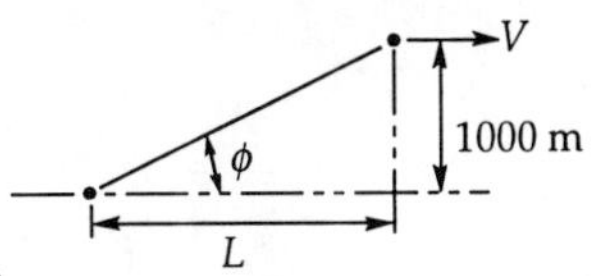

6.82 **d)** Assume $\text{M}_e=0.3$, the maximum if the density is assumed constant (i.e., $\rho_e=0.97\rho_o$). $V_e=\text{M}_ec_e=0.3\sqrt{1.4\times287T_e}.$

$\therefore V_e^2=36.2T_e.$ energy: $0=\frac{V_e^2-V_o^2}{2}+c_p(T_e-T_o).$

$36.2T_e=2\times1000(293-T_e).\quad\therefore T_e=287.8$

$p_o=p_e(T_o/T_e)^{\frac{k}{k-1}}=100\left(\frac{293}{287.8}\right)^{\frac{1.4}{.4}}=106.$

Kinematics and Dynamics of Machinery

by Charles W. Radcliffe and Clark J. Radcliffe

Solutions to Practice Problems

Essay Problem Solutions

7.1

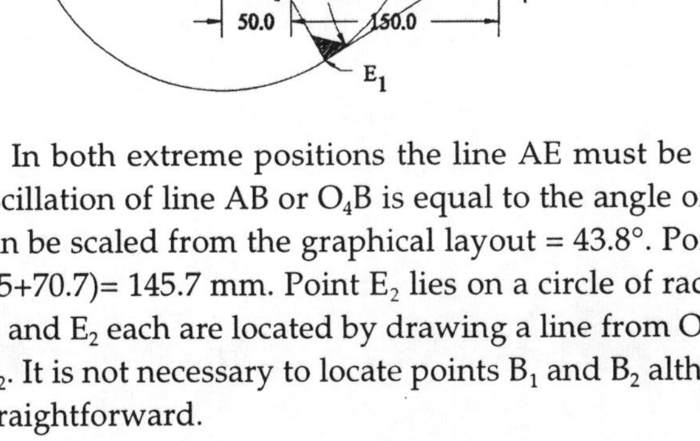

In both extreme positions the line AE must be parallel to O_2A. The angle of oscillation of line AB or O_4B is equal to the angle of oscillation of line EO_4 which can be scaled from the graphical layout = 43.8°. Point E_1 lies on a circle of radius (75+70.7)= 145.7 mm. Point E_2 lies on a circle of radius (75-70.7)= 4.3 mm. Points E_1 and E_2 each are located by drawing a line from O_4 tangent to their circles about O_2. It is not necessary to locate points B_1 and B_2 although completing the layout is straightforward.

7.2

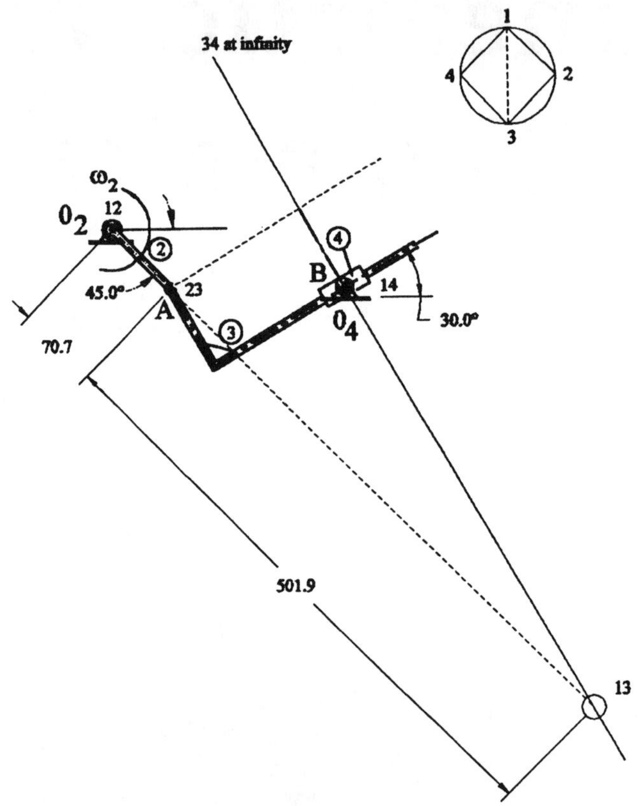

Figure P7.2 Kinematic equivalent mechanism to Figure P7.1

Using the kinematically equivalent mechanism of Figure P7.2 instant center 13 is found using the construction shown above. Since the velocity of instant center 23 is known it is used as the transfer point:

$$V_{23} = \omega_2(12\text{-}23) = (52.36)(70.7) = 3701.9 \text{ mm/sec.}$$
$$\omega_2 = V_{23} / (13\text{-}23) = (3701.9)/(-501.9) = -7.38 \text{ rad/sec}$$

The sign for (13-23) is negative since it is opposite to (12-23).

7.3 The use of the kinematic equivalent shown in Figure P7.2 is justified by noting that both mechanisms provide constraints such that member 3 must have the same angular displacement as member 4 in all positions for the input member 2.

7.4

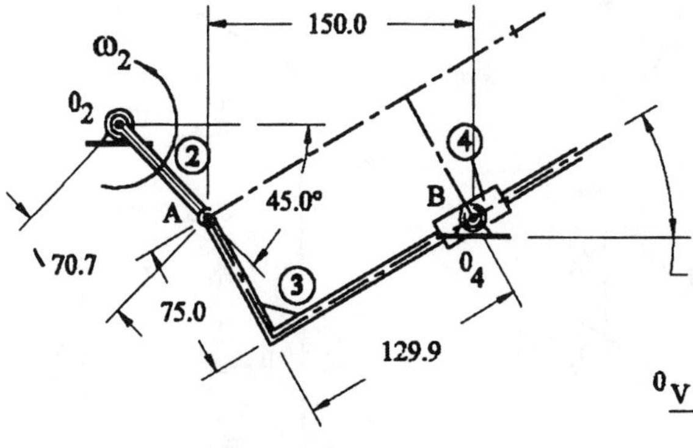

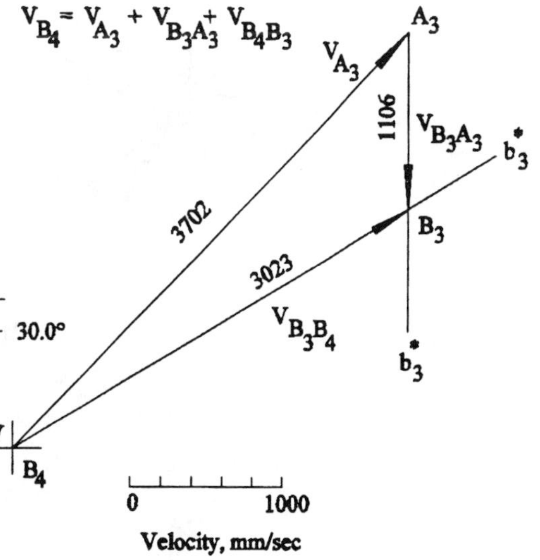

$\omega_3 = 1106/150 = 7.37$ rad/sec

$\omega_4 = \omega_3$

Sliding velocity at point B

$V_{rel} = V_{B_3B_4} = 3023$ mm/sec $= 3.023$ m/sec ↗

The velocity vector equation is written in the form

$$\mathbf{V}_{B4} = \mathbf{V}_{A3} + \mathbf{V}_{B3/A3} + \mathbf{V}_{B4/B3}.$$

Draw $\mathbf{V}_{A3}$ perpendicular to O_2A_2 which locates point A_3 in the velocity diagram. Insert the arrowhead at point A_3. The velocity of point $A_3 = 3702$ mm/sec. The direction of $\mathbf{V}_{B3/A3}$ is perpendicular to the line A_3B_3. Label the line from A_3 in the velocity diagram which contains B_3 as b_3^*. The direction of $\mathbf{V}_{B4/B3}$ is parallel to the slot in the slider. The point B_4 lies at O_V since it has zero velocity. Label the line from B_4 which contains B_3 as a second b_3^* line. The intersection of the b_3^* lines locates point B_3 in the velocity diagram. Add the arrowheads at point B_3 then complete the calculations for ω_2 and sliding velocity $\mathbf{V}_{B3/B4}$.

7.5

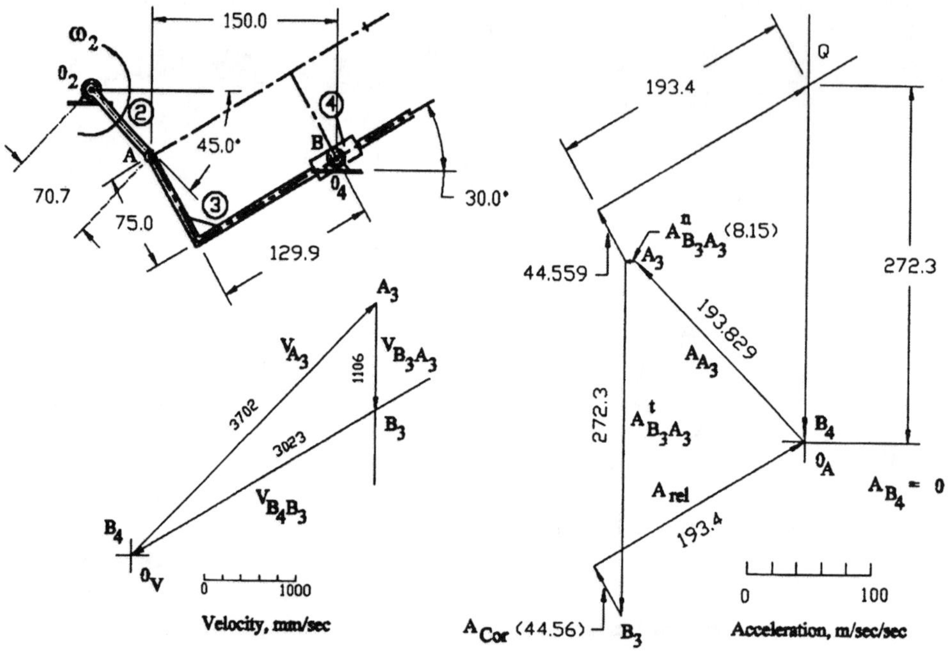

$\alpha_3 = A^t_{B_3A_3} / \overline{A_3B_3} = 272.3 / 0.15 = 1815.33$ rad/sec/sec (Clockwise)

$A_{rel} = 193.4$ m/sec/sec (Scaled from the acceleration diagram)

The velocity vector equation is written in the form

$$\mathbf{V}_{B_4} = \mathbf{V}_{A_3} + \mathbf{V}_{B_3A_3} + \mathbf{V}_{B_4B_3} = \mathbf{V}_{A_3} + \mathbf{V}_{B_3A_3} + \mathbf{V}_{rel}$$

where

$$\omega_3 = V_{B_3A_3} / (A_3B_3) = (1106)/(150) = 7.37 \text{ rad/sec}$$

$$V_{rel} = V_{B_4B_3} = 3023 \text{ mm/sec (as scaled from the velocity polygon)}$$

The acceleration vector equation is written in the form

$$\mathbf{A}_{B_4} = \mathbf{A}_{A_3} + \mathbf{A}_{B_3A_3} + \mathbf{A}_{rel} + \mathbf{A}_{cor}$$

where

$$A_{B_4} = 0$$

$$A_{A_2} = (70.7)(52.36)^2 = 193{,}829 \text{ mm/sec/sec} = 193.8 \text{ m/sec/sec}$$

$A^n_{B_3A_3} = (150)(7.37)^2 = 8148$ mm/sec/sec $= 8.15$ m/sec/sec in a direction parallel to the line A_3B_3

$A^t_{B_3A_3}$ is in a direction perpendicular to the line A_3B_3

$$A_{cor} = 2\ \omega_3 \times V_{B_4B_3} = (2)(7.37)(3023) = 44{,}559 \text{ mm/sec/sec} = 44.56 \text{ m/sec/sec}$$

$A_{rel} = d/dt(V_{rel})$ is in a direction parallel to V_{rel} and $V_{rel} = V_{B_4B_3}$

The direction of the Coriolis component is established by first noting that the relative sliding velocity of point B_4 relative to B_3 is directed downward-left. Rotate the relative velocity vector 90° clockwise (in the direction of ω_3) and it points in the direction of the Coriolis component).

In the solution first draw the vectors in the order $\mathbf{A}_{A3}$, $\mathbf{A}^n_{B3A3}$, $\mathbf{A}_{cor}$, $\mathbf{A}_{rel}$ (direction along slider), A^t_{B3A3} (direction from B_4 at O). The intersection at point Q allows the unknown magnitudes of A_{rel} and A^t_{B3A3} to be scaled from the diagram. To show the vectors in their actual relative positions the diagram is redrawn as shown to the lower-left of point O with point B_3 in its proper location relative to point B_4.

7.6 Instant center 13 is located at point O_4. Instant center 24 is located at point A. Since instant centers 13 and 14 are coincident the angular velocity of members 3 and 4 are equal. The most direct way to find the angular velocity of member 4 is using instant center 24 as a "transfer point":

$V_{24} = (100)(40) = 4000$ mm/sec

$\omega_4 = (4000)/(-100) = -40$ rad/sec

$\omega_3 = \omega_4$

$V_c = 4.32$ m/sec (calculated as shown in the graphical solution below)

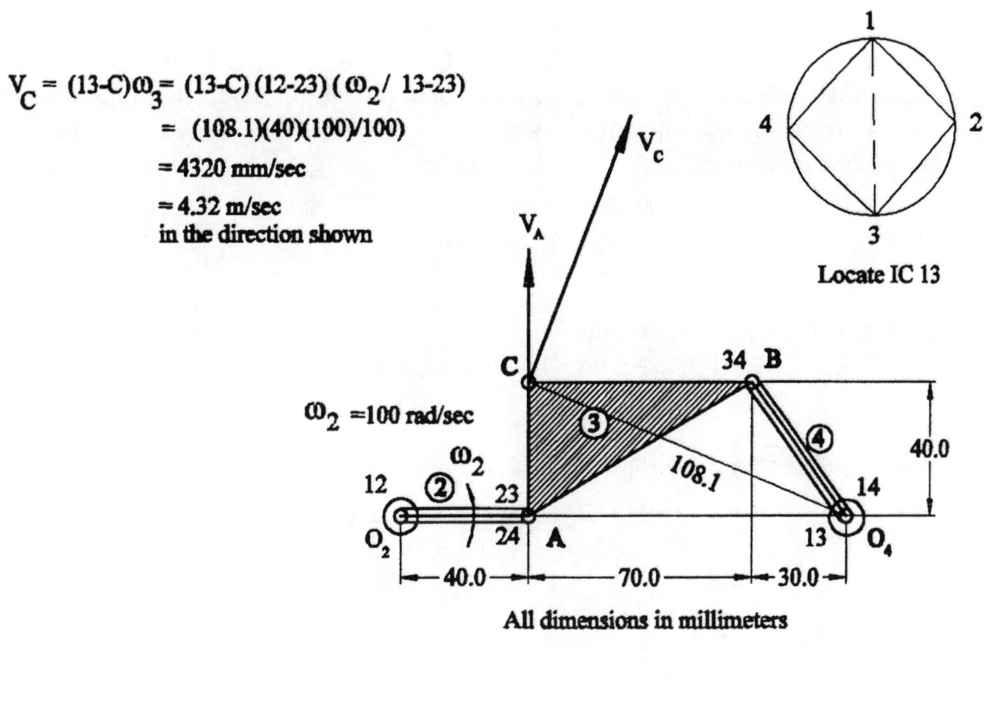

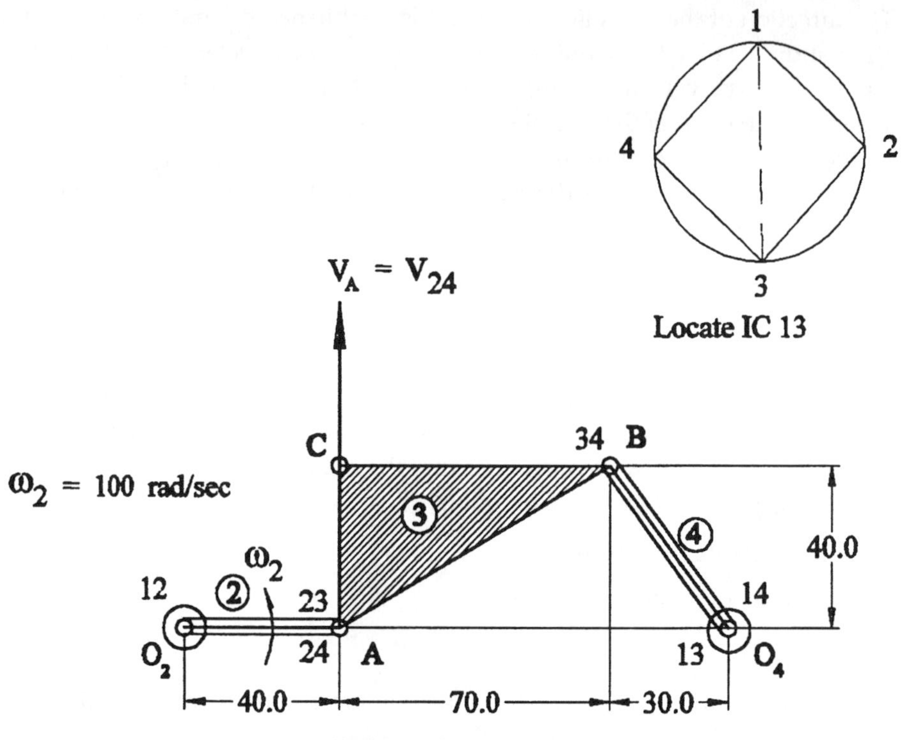

All dimensions in millimeters

Instant center 13 is located at point O_4. Instant center 24 is located at point A. Since instant centers 13 and 14 are coincident the angular velocity of members 3 and 4 are equal. The most direct way to find the angular velocity of member 4 is using instant center 24 as a "transfer point":

$$V_{24} = (100)(40) = 4000 \text{ mm/sec.}$$

$$\omega_4 = (4000)/(-100) = -40 \text{ rad/sec.}$$

$$\omega_3 = \omega_4$$

The velocity of point C becomes

$$V_C = (13 - C)\omega_3 = (108.1)(-40) = 4324 \text{ mm/sec} = 4.32 \text{ m/sec}$$
in the direction shown

7.8 Refer to Chapter 7 for numbered equations

$$\theta_3 = \text{atan}(40/70) = 29.74°, \quad \theta_4 = \text{atan}(40/-30) = 126.87°$$

$$r_{3x} = 70 \; r_{3y} = 40 \; r_{4x} = -30 \; r_{4y} = 40$$

$$\text{det} = r_{3y}r_{4x} - r_{3x}r_{4y} = (40)(-30)-(70)(40) = -1200 - 2800 = -4000$$

Sub into Eq. 7.2.25:

$$\omega_3 = [(0)(-30) + (4000)(40)]/(-4000) = -40 \text{ rad/sec}$$

Sub into Eq. 7.2.26:

$$\omega_4 = [(4000)(40) + (0)(40)]/(-4000) = -40 \text{ rad/sec}$$

$$\dot{\mathbf{c}} = \dot{\mathbf{a}} + \boldsymbol{\omega}_3 \times (\mathbf{c} - \mathbf{a})$$

$$\dot{\mathbf{a}} = \boldsymbol{\omega}_2 \times \mathbf{r}_2 = \begin{vmatrix} \mathbf{i} & \mathbf{j} & \mathbf{k} \\ 0 & 0 & \omega_2 \\ 40 & 0 & 0 \end{vmatrix} = \mathbf{j}(40)(100) = 4000\mathbf{j}$$

$$\boldsymbol{\omega}_3 \times (\mathbf{c}-\mathbf{a}) = \begin{vmatrix} \mathbf{i} & \mathbf{j} & \mathbf{k} \\ 0 & 0 & \omega_3 \\ 0 & -40 & 0 \end{vmatrix} = -\mathbf{i}(40)(-40) = 1600\mathbf{i}$$

Hence $\dot{\mathbf{c}} = 1600\mathbf{i} + 4000\mathbf{j}$ = 4308 mm/sec at an angle of 68° from the x-axis.

7.9 Velocity Analysis

Refer to Chapter 7 for numbered equations:

$$\theta_3 = \text{atan}(40/70) = 29.74°, \quad \theta_4 = \text{atan}(40/\text{-}30) = 126.87°$$

$$r_{3x} = 70 \; r_{3y} = 40 \; r_{4x} = -30 \; r_{4y} = 40$$

$$\det = r_{3y}r_{4x} - r_{3x}r_{4y} = (40)(\text{-}30)\text{-}(70)(40) = -1200 - 2800 = -4000$$

Sub into Eq. 7.2.25:

$$\omega_3 = [(0)(-30) + (4000)(40)]/(-4000) = -40 \text{ rad/sec}$$

Sub into Eq. 7.2.26:

$$\omega_4 = [(4000)(40) + (0)(40)]/(-4000) = -40 \text{ rad/sec}$$

$$\dot{\mathbf{c}} = \dot{\mathbf{a}} + \boldsymbol{\omega}_3 \times (\mathbf{c}-\mathbf{a})$$

$$\dot{\mathbf{a}} = \boldsymbol{\omega}_2 \times \mathbf{r}_2 = \begin{vmatrix} \mathbf{i} & \mathbf{j} & \mathbf{k} \\ 0 & 0 & \omega_2 \\ 40 & 0 & 0 \end{vmatrix} = \mathbf{j}(40)(100) = 4000\mathbf{j}$$

$$\boldsymbol{\omega}_3 \times (\mathbf{c}-\mathbf{a}) = \begin{vmatrix} \mathbf{i} & \mathbf{j} & \mathbf{k} \\ 0 & 0 & \omega_3 \\ 0 & -40 & 0 \end{vmatrix} = -\mathbf{i}(40)(-40) = 1600\mathbf{i}$$

Hence $\dot{\mathbf{c}} = 1600\mathbf{i} + 4000\mathbf{j}$ = 4308 mm/sec at an angle of 68 deg from the x-axis.

Acceleration Analysis

Refer to Chapter 7 for numbered equations

$$\ddot{a}_x = -r_{2x}\omega_2^2 = \text{-}(40)(100)^2 = \text{-}400{,}000 \text{ mm/sec/sec}$$

$$\ddot{a}_y = -r_{2y}\omega_2^2 = \text{-}(0)(100)^2 = 0$$

$$\theta_3 = \text{atan}(40/70) = 29.74°, \quad \theta_4 = \text{atan}(40/\text{-}30) = 126.87°$$

$$r_{3x} = 70 \; r_{3y} = 40 \; r_{4x} = -30 \; r_{4y} = 40$$

$$\det = r_{3y}r_{4x} - r_{3x}r_{4y} = (40)(\text{-}30)\text{-}(70)(40) = -1200 - 2800 = -4000$$

Sub into Eq. 7.2.25:

$$\omega_3 = [(0)(-30) + (4000)(40)]/(-4000) = -40 \text{ rad/sec}$$

Sub into Eq. 7.2.26:

$$\omega_4 = [(4000)(40) + (0)(40)]/(-4000) = -40 \text{ rad/sec}$$

Sub into the first of Eq. 7.2.34:

$$\alpha_3 = \frac{\left[-400{,}000 - (70)(40^2) + (-30)(40^2)\right](-30) + \left[0 - (40)(40^2) + (40)(40^2)\right](40)}{-4000}$$

$$= \frac{(-400{,}000 - 112{,}000 - 48{,}000)(-30) + (0)}{-4{,}000} = -4{,}200 \text{ rad/sec/sec}$$

Sub into the second of Eq. 7.2.34:

$$\alpha_4 = \frac{\left[0-(40)(40^2)+(40)(40^2)\right](40)+\left[-400,000-(70)(40^2)+(-30)(40^2)\right](70)}{-4000}$$

$$= \frac{(0)+(-400,000-112,000-48,000)(70)}{-4000} = 9800 \text{ rad/sec/sec}$$

$$\ddot{\mathbf{c}} = \ddot{\mathbf{a}} + \boldsymbol{\alpha}_3 \times \mathbf{s}_C + \boldsymbol{\omega}_3 \times (\boldsymbol{\omega}_3 \times \mathbf{s}_C)$$

$$\ddot{c}_x = \ddot{a}_x - s_{cy}\alpha_3 - s_{cx}\omega_3^2 = -(40)(100^2)-(40)(-4,200)-(0)(-40^2) = -232,000 \text{ mm/sec/sec}$$

$$= -232 \text{ m/sec/sec}$$

$$\ddot{c}_y = \ddot{a}_y + s_{cx}\alpha_3 - s_{cy}\omega_3^2 = (0)+(0)-(40)(40^2) = -64,000 \text{ mm/sec/sec}$$

$$= -64 \text{ m/sec/sec}$$

The direction of the acceleration for point C is atan 2 (–64, –232)= 195.4°.

7.10 Solution using the formula method

In order for gears 4 and 5 to mesh the tooth numbers must be proportional to their radii. Since the radius of the planet gear 4 equals 2 inches, the radius of the ring gear 5 equals 14 inches.

(a) The number of teeth on the ring gear = (14/2)(20) = 140 teeth

Using Eq. 7.4.10 (The Formula Method) the driver gear is the fixed ring gear with an angular velocity = 0. The driven gear is the sun gear:

$$VR = \mp\frac{\text{Product of number of teeth on driver gears}}{\text{Product ofnumber of teeth on driven gears}} = \frac{\omega_{\text{Last}} - \omega_{\text{Arm}}}{\omega_{\text{First}} - \omega_{\text{Arm}}}$$

$$\therefore \quad -\frac{140}{100} = \frac{\omega_2 - 100}{0-100}$$

from which $\omega_2 = -(100)(-1.4)+100 = 240$ rpm

Alternate Solution 7.10 —Using the method of relative instant centers

The transfer point, instant center 24, is the pitch point (contact point) for gears 2 and 4. The velocity of the center of the planet gear 4 = (12)(100)(π/30) = (1200)(π/30). Therefore the velocity of the transfer point 24 = (2400)(π/30). With the velocity of the transfer point 24 known, the angular velocity of the sun gear 2 can be calculated as ω_2 = (2400π/30)/10 = (240)(π/30) rad/sec. Converting back to rpm gives ω_2 = 240 rpm.

7.11 Since the angular velocity ratio is greater than 10:1 a double reduction compound set of spur gears is indicated. Assume a gearset ratio of 5 for the first set and a gearset ratio of 4 for the second set. For a 20 degree pressure angle assume 14 teeth on the driving gears 1 and 3 of the compound gearset. This would require 84 teeth on driven gear 2 and 70 teeth on driven gear 4. Checking the overall gear train ratio *VR* = ((84)(70))/((14)(14))= 20.

For an assumed diametral pitch = 5 the center distance for the first gearset = 84/5 = 16.8 inches. The center distance for the second gearset = 70/5 = 14.0 inches. This gives an overall center distance = 30.8 inches which exceeds the specification.

Since $(84/p_d) + (70/p_d)$ must be less than 10, p_d must be greater than (154/10).

Try a diametral pitch = 16: d_1= 84/16=5.25 inches, d_2= 70/16 = 4.375 inches. $d_1 + d_2$ = 9.625 inches The circular pitch for the first gearset = $\pi d/N$ = (3.1416)(5.25)/(84) = 0.196 inches. For the second gearset the circular pitch = (3.1416)(4.375)/(70 = 0.196 as expected.

Solution sketch for 7.12-7.21:

Multiple Choice Problem Solutions

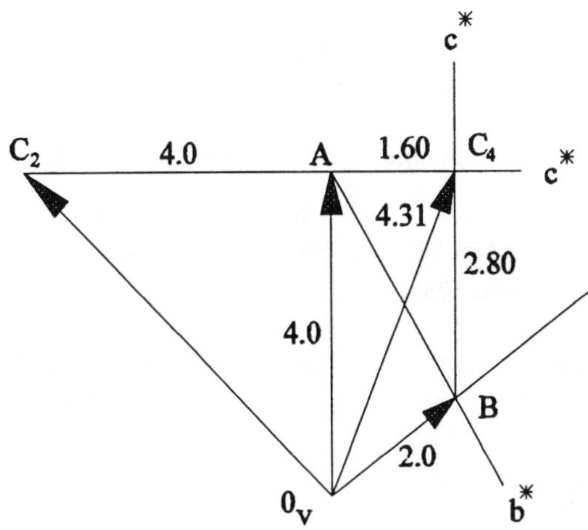

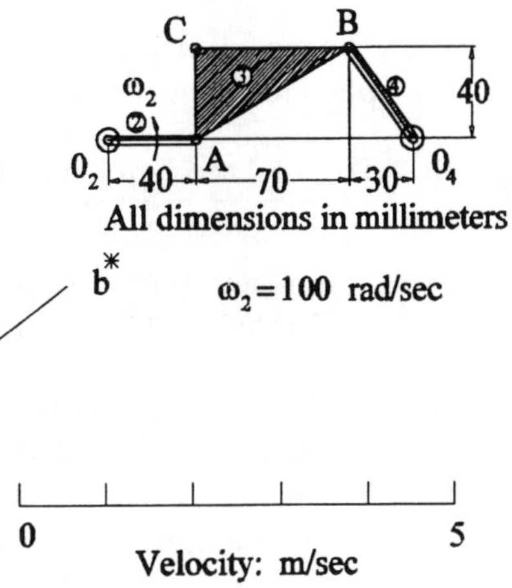

7.12 **c)** The velocity of point C is 4.31 m/sec

7.13 **e)** The angular velocity of member 3 = 1.60/.040 = 40 rad/sec CW

7.14 **b)** The angular velocity of member 4 = 2.00/.050 = 40 rad/sec CW

7.15 **b)** The distance 14-C = $\sqrt{(70+30)^2+(40)^2}$ = 108. V_C= (0.108)(40)=4.31 m/sec

7.16 **a)** $\omega_3 = V_{23}/(13\text{-}23) = (12\text{-}23)\omega_2/(13\text{-}23)$ = (0.040)(100)/(0.100) = 40 rad/sec CW

7.17 **a)** $\omega_4 = V_{24}/(14\text{-}24) = (12\text{-}24)\omega_2/(14\text{-}24)$ = (0.040)(100)/(0.100) = 40 rad/sec CW

7.18 **e)** The velocity of points C_3 and C_4 are equal hence the relative velocity must be zero.

7.19 **e)** $\omega_{rel} = \omega_3 - \omega_4 = -40 - (-40) = 0$

7.20 **d)** The relative velocity $V_{C2} - V_{C4}$ is seen in the velocity vector diagram.

7.21 **d)** $\omega_{rel} = \omega_2 - \omega_4 = 100 - (-40) = 140$ rad/sec

7.22 **c)** The stretched length of the spring = $\sqrt{(6.0-4.361)^2+(4.0-1.340)^2}$ =3.124 at an angle atan[(4.0 - 1.340)/(6.0 - 4.361)] = 58.36 deg. The tension in the spring is (10.0)(3.124-2.0) = 11.24 lbs.

7.23 **b)** The velocity of point B is $\sqrt{-18.61^2+42.89^2}$ = 46.75 in./sec This velocity vector is directed at an angle atan((42.89)/(-18.61)) = 113.46° from the positive x-axis. The axis of the damper is directed at an angle atan[(1.0-(-0.995)/(1.0-1.706)] = 109.49°. Therefore the axial relative velocity in the damper = (46.75)cos(113.46-109.49) = 46.64 in./sec. The compressive force in the damper = (1.20)(46.64) = 55.97 lbs.

7.24 **b)** As a result of the input torque about O_2 the force at A has a transverse component therefore the forces at O_2 and A cannot be colinear.

7.25 **a)** Forces are applied to member 3 at three distinct points.

7.26 **d)** The mass is negligible, hence no inertia effects, and no forces are applied to member 4 between points O_4 and B.

7.27 **d)** The total acceleration of the mass center G_3 for member 3 = $\sqrt{(-1095)^2 + (-300.5)^2}$ = 1135 in./sec.^{2}. The direction of the acceleration of G_3 = atan[(-300.5)/(-1095)] = 195.35°. The effective force = (0.03)(1135)= 34 lbs. at an angle = 195 deg.

7.28 **b)** Since the angular acceleration of member 3 is positive (counter-clockwise), the effective force must pass above the center of mass of member 3 with an eccentric offset = (0.003)(1506)/(34.06) = 0.0013 in. (Not large since the angular acceleration is small.)

7.29 **c)** See Figure P7.10 below

7.30 **d)** See Figure P7.10 below

7.31 **b)** See Figure P7.10 below

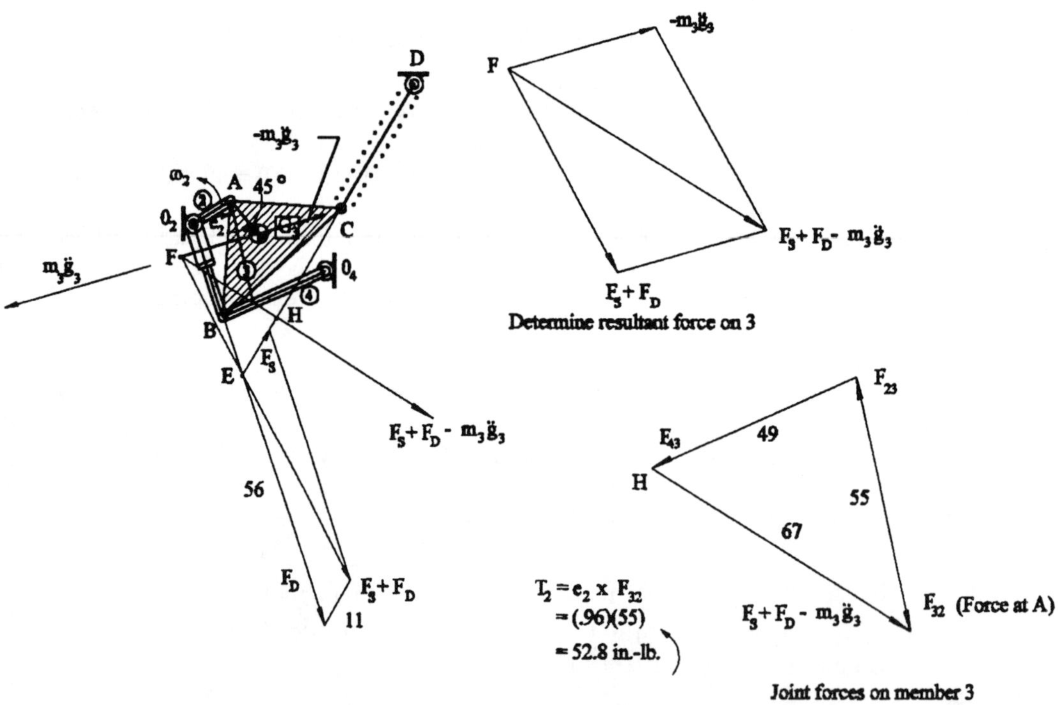

Figure P7.10 Graphical solution Problems 7.22-31

The graphical solution procedure is as follows:

(a) Begin by placing the reversed effective force $-m_3\ddot{g}_3$ on the diagram in a direction opposite but collinear with the effective force $m_3\ddot{g}_3$.

(b) Extend the lines of action of the spring and damper forces. At the point of intersection E determine the vector sum $\mathbf{F_S} + \mathbf{F_D}$.

(c) Extend the line of action of the sum $\mathbf{F_S} + \mathbf{F_D}$ from (b) back to the line of the reversed effective force (inertia force) and add the inertia force to the sum $\mathbf{F_S} + \mathbf{F_D}$ at the intersection point F.

(d) Locate the intersection point H of the force vector found in (c) and the known direction of $\mathbf{F}_{43}$.

(e) The force $\mathbf{F}_{23}$ from point A intersects at the point H found in (d).

(f) Solve the vector triangle of the forces acting on member 3 , with known directions for $\mathbf{F}_{43}$ and $\mathbf{F}_{23}$, to obtain their magnitudes.

(g) With $\mathbf{F}_{32} = -\mathbf{F}_{23}$ known, measure the offset $\mathbf{e}_2$ from O_2 and compute T_2 = (0.96)(55) = 52.8 in.-lb.

7.32 **d)** *VR*= (-18/24)(20/50)= 0.3. Output rpm= (0.3)(1750)= 525 rpm

7.33 **b)** The diameter of gear 3 = 20/5 = 4.0 in. The diameter of gear 4 = 10 - (2.0 + 5.0) = 3.0 The number of teeth on gear 4 = (5)(3) = 15

7.34 **e)** Gear 4 reverses the output direction and sets the center distance = 10.0

7.35 **c)** The output speed is independent of efficiency. Ans same as Prob 7.32

7.36 **d)** Torque= (.90) 5252 Hp/rpm = (.90)(5252)(25)/(500) = 236.3 ft-lb

7.37 **b)** Gears with the smallest diametral pitch.

7.38 **b)** Gears with the smallest diametral pitch.

7.39 **c)** Interchangeable gears must have the same diametral pitch (or module) and the same addendum/dedendum.

7.40 **c)** Since the limit per stage is 10:1 a three-stage gear set would be required.

7.41 **b)** The pressure angle increases, the efficiency decreases, the velocity ratio is unchanged.

Stress Analysis & Structures

by Clark J. Radcliffe and Charles W. Radcliffe

Solutions to Practice Problems

Essay Problem Solutions

8.1 a) Draw a free-body diagram of half the linkage cut through plane A-A.

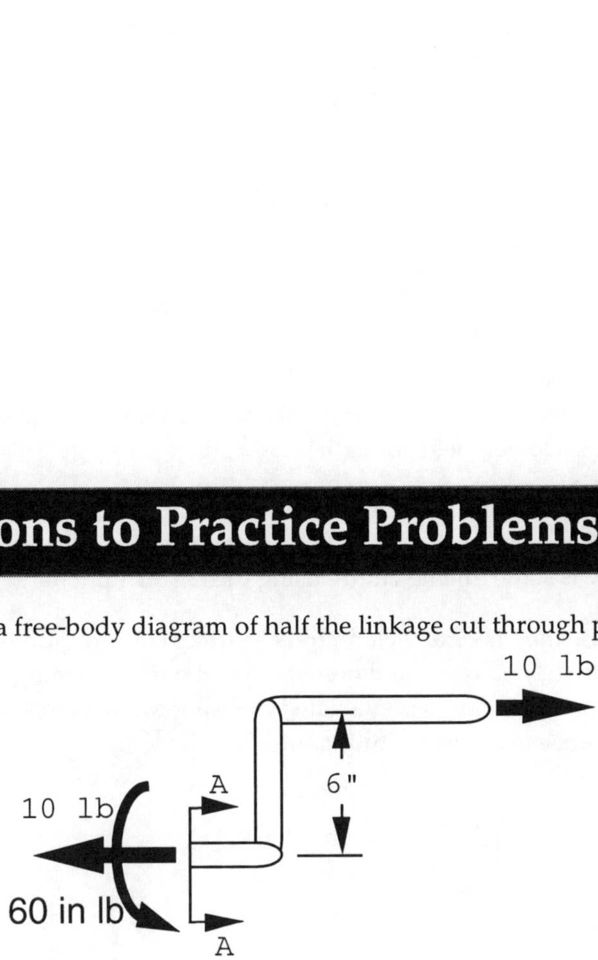

The force and moment reactions at plane A-A are a 10 lb tension force and a 60 in-lb moment. The stress in the section includes an axial normal stress σ_t from the tensile force and a normal bending stress σ_b from the bending moment. On the top edge of the linkage, the total plane A-A normal stress $\sigma = \sigma_t + \sigma_b = \frac{F}{A} + \frac{My}{I}$ is the sum of the tensile and bending stress components. Because there are no other stress components at this section, the von-Mises stress σ' equals the maximum normal stress at this point where $y = d/2$, $A = \pi d^2/4$, $I = \pi d^4/64$ and $d = 0.5$ in:

$$\sigma' = \sigma_t + \sigma_b = \frac{F}{A} + \frac{My}{I} = \frac{4F}{\pi d^2} + \frac{32M}{\pi d^3} = 4.94 \times 10^3 \text{ lb/in}^2$$

b) For 500 cycles, both "low cycle fatigue" and "yield" failures are analyzed, where $S_{yt} = 39.5$ kpsi and $S_{ut} = 72$ kpsi from Table 8.2. The material's low cycle fatigue strength at 500 cycles is determined using interpolation and Figure 8.17 as

$$S_f = \left[1 - 0.1\left(\frac{\log(500)}{\log(10^3)}\right)\right] S_{ut} = [0.91]72 \text{ kpsi} = 65.5 \text{ kpsi}$$

which is larger than the materials' yield strength. Using the lower material yield strength, the factor of safety for completely reversed loading is $n = S_{yt}/\sigma' = 39.5 \text{ kpsi}/4.94 \text{ kpsi} = 8.0$. Typically, ductile materials fail in yield at loads smaller than their low cycle fatigue strengths.

c) For 100,000 cycles, "high cycle" fatigue strength will be evaluated. Using interpolation and Figure 8.17,

$$S_f = \left[0.9 - 0.4\left(\frac{\log(10^5) - \log(10^3)}{\log(10^6) - \log(10^3)}\right)\right] S_{ut} = [0.633]72 \text{ kpsi} = 45.6 \text{ kpsi}$$

and the factor of safety at 100,000 cycles remains $n = 8.0$ since the material has a lower yield strength.

8.2 a) This axle's material has more than 3% elongation at failure so is considered ductile and *the axle will fail in material yield.* The appropriate failure theory is static ductile failure using the Distortion Energy Theory.

b) Because the material yields, stress at stress concentrations remain approximately constant and do not affect the failure strength of the axle. The von Mises stress is evaluated by using Mohr's circle to find the principal stresses and substituting into Eq.8.4.2:

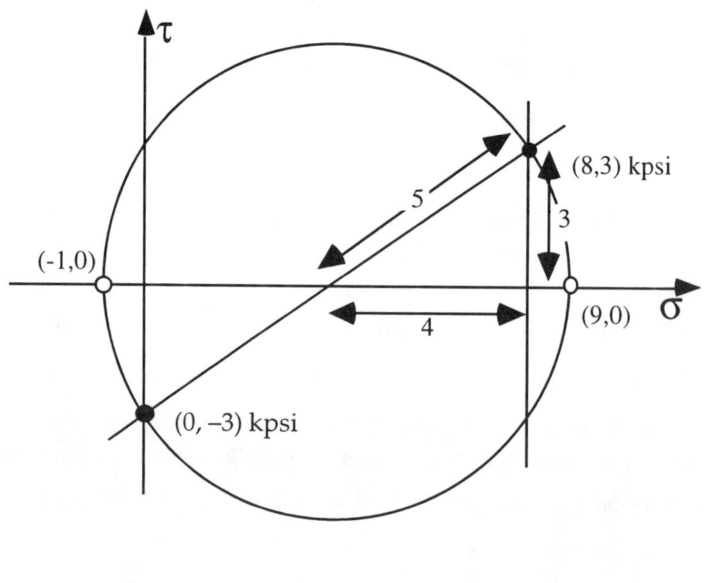

$$\sigma' = \sqrt{\frac{(9-(-1))^2 + ((-1)-0)^2 + (0-9)^2}{2}} = \sqrt{\frac{100+1+81}{2}} = \sqrt{91} = 9.54 \text{ kpsi}$$

The factor of safety for static ductile failure is $n = S_{ut}/\sigma' = 50/9.54 = 5.24$

8.3 a) This axle has the same geometry and loading as in Problem 8.2, however, the material has different compressive and tensile stresses indicating a brittle material. For this reason, *the material failure is evaluated using the Modified-Mohr failure theory.* Both load stresses must be increased by appropriate stress concentration factors for $r/d = 0.1$ and $D/d = 1.5$.

b) The torsional stress concentration $K_t = 1.5$ from Figure 8.10 gives a torsional stress $\tau = K_t\tau_0 = 1.5(3) = 4.5$ kpsi(cw). The bending stress concentration $K_t = 1.7$ from Figure 8.11 gives a bending stress $\sigma_b = K_t\sigma_0 =$ 1.7(8)= 13.6 kpsi. The principal stresses are found from a Mohr's circle analysis (graphic not reproduced here) to be

$$\sigma_1, \sigma_2 = \left(\frac{13.6}{2}\right) \pm \sqrt{\left(\frac{13.6}{2}\right)^2 + (4.5)^2} = 15.0, -1.35$$

Reproducing the fourth quadrant of Figure 8.16 the load line is determined by the combination of a $\sigma_1 = 15.0$ kpsi and $\sigma_2 = -1.35$ kpsi and crosses the Modified-Mohr failure criteria at $S_{ut} = 50$ kpsi with a slope of -1.35/15 (see the figure that follows). Because the load line slope is less than 1, the factor of safety $n = S_{ut}/\sigma_1 = 50/15 = 3.33$ is determined by the maximum normal stress.

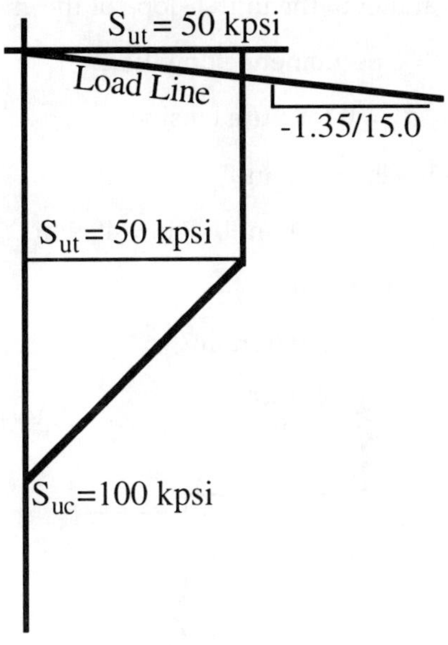

8.4 a) The same axle geometry as Problem 8.2 but constructed of mild steel with a machined finish and subjected to a rotating bending stress has a life governed by fatigue failure. The endurance limit Eq.8.4.6:
$S_e = k_a k_b k_c k_d k_e S_e' = (0.958)(0.944)(1)(1)(.588)(0.5 \times 50)$ kpsi $= 13.3$ kpsi

where $k_a = 2.7(50)^{-0.265} = 0.958$ (from Table 8.3)

$$k_b = \left(\frac{0.5}{0.3}\right)^{-0.1133} = 0.944 \text{ (from Eq.8.4.8a)}$$
$$k_c = 1.0 \text{ (from Eq.8.4.9)}$$
$$k_d = 1.0 \text{ (from Eq.8.4.10 because no temperature specified)}$$
$$k_e = 1/K_f = 1/1.7 = 0.588 \text{ (from Eq.8.4.11)}$$

b) At N=10[5] cycles, use the unmodified load stress $\sigma_b = 8$ kpsi since stress concentration K_f was included in the endurance limit calculation. Compute the "low cycle" strength $0.9S_t$ without stress concentration K_f because the material is ductile. Reproducing Fig. 8.17, and interpolating for the modified endurance limit S_e:

$$S_f = 0.9S_{ut} - \left(\frac{(5-3)}{(6-3)}\right)(0.9S_{ut} - S_e) = 0.9(50) - \left(\frac{2}{3}\right)(0.9(50) - 13.3) = 23.9 \text{ kpsi}$$

The factor of safety

$$n = S/\sigma = 23.9/8 = 2.99$$

Multiple Choice Problem Solutions

8.5 **d)** $E = 10$ MPa/0.01 = 1000 MPa

8.6 **a)** Graph at $\varepsilon = 0.002$ parallel to slope 1000 MPa-mm/mm yields 25 MPa

8.7 **b)** $S = \sigma = 37$ MPa, the peak of the stress strain curve, then the load supported falls.

8.8 **c)** 6%, extend a line parallel to the initial slope of the curve to the point of failure.

8.9 **a)** ductile, more than 3% permanent elongation.

8.10 **a)** $\sigma_1 = 35$ MPa, $\sigma_2 = \sigma_3 = 0$ for pure tension.

8.11 **c)** (6 in/2) × (200 lb - 100 lb) = 300 in-lb

8.12 **b)** F(7 in) = 300 in-lb gives F = 300 in-lb/7 in = 42.9 lb

8.13 **c)** 300 in-lb neglecting bearing drag.

8.14 **a)** Draw a plan-view free-body-diagram:

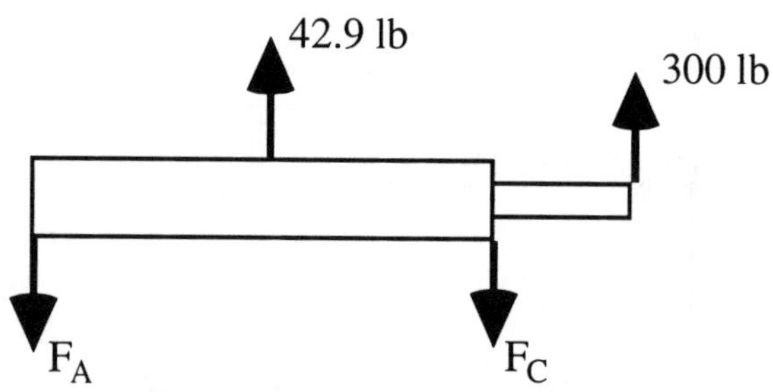

$$F_A(26 \text{ in}) + 300 \text{ lb}(6 \text{ in}) = 42.9(13 \text{ in})$$
$$\Rightarrow F_A = [42.9(13 \text{ in}) - 300 \text{ lb}(6 \text{ in})]/(26 \text{ in})$$
$$\Rightarrow F_A = -47.8 \text{ lb}$$

8.15 **d)** From the free-body-diagram

$$F_C(26 \text{ in}) = 300 \text{ lb}(19 \text{ in}) + 42.9(13 \text{ in})$$

$$\Rightarrow F_C = [300 \text{ lb}(32 \text{ in}) + 42.9(13 \text{ in})]/(26 \text{ in})$$

$$\Rightarrow F_C = 390.7 \text{ lb}$$

8.16 **c)** $M = (300 \text{ lb})(6 \text{ in}) = 1800 \text{ in-lb}$

8.17 **a)** $S = \sigma = My/I \Rightarrow 26 \text{ kpsi} = [1800 \text{ in lb}(d/2)]/[\pi d^4/64]$ (ductile failure)

$\Rightarrow d^3 = [1800 \text{ in lb}(32)]/[\pi(26 \text{ kpsi})] = 0.71 \text{ in}^3$

$\Rightarrow d = 0.89 \text{ in} \Rightarrow d = 1.0$ in with increase to next largest standard size

8.18 **b)** $S_e = k_a k_b k_c k_d k_e S_e' = (0.91)(.81)(1)(1)(0.59)[0.5(47 \text{ kpsi})] = 10.2$ kpsi using Eq.8.4.6 where $k_a = 0.91$(HR finish), $k_b = 0.81$ @ d= 2 in (0.87 @1 in), $k_c =$ 1.0 (bending), $k_d = 1.0$ (temperature unspecified), and $k_e = 1/1.7 = 0.59$ (Fig 8.11) then $S = \sigma = My/I \Rightarrow 10.2 \text{ kpsi} = [1800(d/2)]/[\pi d^4/64] \Rightarrow$ $d = 1.22 \text{ in} \Rightarrow 1.25 \text{ in}$

Vibration

by Clark J. Radcliffe and Charles W. Radcliffe

Solutions to Practice Problems

Essay Problem Solutions

9.1

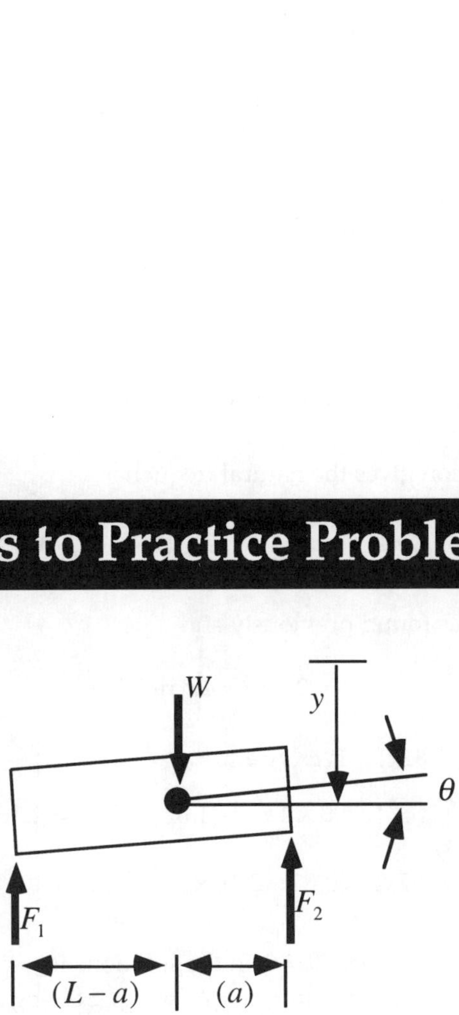

a) First sketch the free-body diagram. Here the wheelbase L determines the distance between the suspension forces while the distance a determines the position of the vehicle Center-of-Gravity (CG is traditionally measured back from the front of the vehicle). Here the vertical motion y of the CG combined with rotation θ yields spring forces $F_1 = k_1\Delta_1 = k_1\left(y + (L-a)\sin\theta\right)$ and $F_2 = k_2\Delta_2 = k_2\left(y - a\sin\theta\right)$.

Now find the vertical and rotational equations of motion:

$$M\ddot{y} = W - F_1 - F_2 = W - k_1\left(y + (L-a)\sin\theta\right) - k_2\left(y - a\sin\theta\right)$$
$$J\ddot{\theta} = F_1(L-a) - F_2 a = k_1\left(y + (L-a)\sin\theta\right)(L-a) - k_2\left(y - a\sin\theta\right)a$$

For the springs to provide no net rotation on the vehicle, we require $\ddot{\theta} = \dot{\theta} = \theta = 0$. When these conditions are substituted into the second equation of motion we derive

$$0 = k_1 y(L-a) - k_2 ya \quad \text{or} \quad k_1 = \frac{a}{(L-a)} k_2$$

From the same rotation equation we get the condition

$$0 = F_1(L-a) - F_2 a = 0.45W(L-a) - 0.55Wa$$

which yields $a = 0.45L$, so that

$$k_1 = \frac{a}{(L-a)} k_2 = \frac{0.45L}{(L-0.45L)} k_2 = \frac{0.45L}{0.55L} k_2 = 0.82k_2$$

Substituting the constraint on rotation and the relation between stiffnesses into the vertical equation of motion yields

$$M\ddot{y} = W - k_1 y - k_2 y = W - (k_1 + k_2)y$$

and rearranging

$$M\ddot{y} + (k_1 + k_2)y = W$$

This equation of motion gives the natural frequency

$$\omega_n = \sqrt{(k_1 + k_2)/M} = \sqrt{(k_1 + k_2)g/W}$$

where $k_1 = 0.82k_2$ was found previously and

$$\omega_n = 2\pi \times 2 = 4\pi \text{ rad / sec}$$

Solving for k_2 yields $1.82k_2 = W\omega_n^2/g$ or

$$k_2 = W\omega_n^2/1.82g = \left(3200 \times 4\pi^2\right)/(1.82 \times 32.2) = 2.16 \times 10^3 \text{ lb / ft}$$

$$k_1 = 0.82k_2 = 0.82 \times 2.16 \times 10^3 = 1.76 \times 10^3 \text{ lb/ft}$$

b) Because we have constrained $\theta = 0$ in the solutions for k_1 and k_2, the deflections of both springs will be the same $\Delta = \Delta_1 = \Delta_2$. Solving for deflection Δ_2 of spring k_2 supporting 55% of the vehicle weight

$$\Delta = \Delta_1 = \Delta_2 = 0.55W/k_2 = 0.55 \times 3200/2.16 \times 10^3 = 0.81 \text{ ft} = 9.8 \text{ in.}$$

9.2

Using Section 9.8, the absorber natural frequency must match the frequency of the pump shaking force so that

$$\omega_n = 2\pi \times 30 = 188 \text{ rad / sec}$$

Using Eq.9.8.4, the absorber stiffness is found from the displacement constraint

$$k_2 = F_2/X_2 = 100/0.01 = 10\,000 \text{ N/m}$$

Using Eq.9.8.2, the absorber mass should be

$$m_2 = k_2/\omega_n^2 = 1000/188^2 = 0.283 \text{ kg}$$

Multiple Choice Problem Solutions

9.3 **e)** The four isolator stiffness are in parallel and each provide 1/4 of the mounting system's effective stiffness

$$k = 4 \times 100 \text{ lb / in} = 400 \text{ lb / in}$$

9.4 **b)** The isolator each support 1/4 of the weight of the pump and

$$\Delta = (500 \text{ lb}/4)/\ (100 \text{ lb/in}) = 1.25 \text{ in}$$

9.5 **d)** Each isolator provides 1/4 of the total stiffness of the mounts applied to the total mass of the pump. The natural frequency

$$\omega_n = \sqrt{(4k)/(W/g)} = \sqrt{(4 \times 100 \times 12)/(500/32.2)} = 17.6 \text{ rad / sec} = 2.80 \text{ Hz}$$

9.6 **b)** The frequency response of the pump (9.4.2) is dominated by the inertia of the pump at high rotational frequency

$$X = \left(me\omega^2\right)\Big/\sqrt{\left(k - M\omega^2\right)^2 + (c\omega)^2} \approx me/M = \left((mg)/Mg\right)e = (1.0/500.)5 \text{ in} = 0.01 \text{ in}$$

9.7 **c)** The non-dimensional frequency response (Eq.9.4.4) of the pump at its natural frequency $(\omega/\omega_n) = 1$

$$\left(\frac{M}{m}\right)\frac{X}{e} = \frac{\left(\dfrac{\omega}{\omega_n}\right)}{\sqrt{\left(1-\left(\dfrac{\omega}{\omega_n}\right)^2\right)^2 + \left(2\zeta\left(\dfrac{\omega}{\omega_n}\right)\right)^2}} = \frac{1}{\sqrt{\left(1-1^2\right)^2 + (2\zeta)^2}} = \frac{1}{2\zeta} \quad \text{so that}$$

$$\zeta = \frac{1}{2}\left(\frac{m}{M}\right)\frac{e}{X} = \frac{1}{2}\left(\frac{1.0}{500}\right)\frac{5.0 \text{ in}}{1.0 \text{ in}} = \frac{1}{200} = 0.05 = 5\%$$

9.8 **a)** The frequency response (Eq.9.4.2) of the pump at the operating rotational frequency is

$$\omega = 2\pi f = 2\pi(30 \text{ cps}) = 188.5 \text{ rad/sec}$$

with damping coefficient (Eq.9.2.2)

$$c = 2\zeta\sqrt{km} = 2 \times 0.05\sqrt{(12 \times 400) \times (500/32.2)} = 1.76 \text{ lb} \cdot \text{sec}^2 \text{ / ft}$$

is

$$X = \left(me\omega^2\right) \Big/ \sqrt{\left(k - M\omega^2\right)^2 + (c\omega)^2}$$

$$= \left[(1/32.2)(5/12)188.5^2\right] \Big/ \sqrt{\left[(4 \times 100) - (500/32.2)188.5^2\right]^2 + (1.76 \times 188.5)^2}$$

$$= [459.8] \Big/ \left[551 \times 10^3\right] = 0.834 \times 10^{-3} \text{ ft} = 0.01 \text{ in}$$

Note that the effect of damping is negligible at a damping ratio of 5%.

9.9 **c)** Assuming the surface does not move, the transmitted force

$$F = kx + c\dot{x} = kX + j\omega cX = (k + j\omega c)X$$

$$= [(400)12 + j(188.5)(1.76)](0.01/12)$$

$$= [(4800) + j(332)]\left(0.834 \times 10^{-3}\right)$$

$$= [4.00 + j0.277] = \sqrt{4.00^2 + 0.277^2} \text{ lb at } \theta = \tan^{-1}(0.277/4)$$

$$= 4.01 \text{ lb at } \theta = 3.96 \text{ deg}$$

9.10 **c)** From the graph, the displacement amplitude varies from 1 mm at 0.1 cyc/sec to 10 mm at 10 cyc/sec.

9.11 **c)** Velocity amplitude is proportional to the product of frequency and displacement. From the graph, the largest velocity must therefore occur at f = 10 Hz = 10 cyc/sec by inspection.

9.12 **e)** The velocity magnitudes ($V = \omega X = 2\pi f X$) at the corner points in the graph are

V = 3.77, 37.7,628. and 314. mm/sec at

frequencies f = 0.1, 1.0, 10, and 50 Hz, respectively.

The largest magnitude is V = 628 mm/sec at f = 10 Hz.

9.13 **d)** Acceleration amplitude is proportional to the product of frequency-squared and displacement. From the graph, the largest velocity must therefore occur at f = 50 Hz = 50 cyc/sec by inspection.

9.14 **c)** The acceleration magnitudes ($A = \omega^2 X = 4\pi^2 f^2 X$) at the corner points in the graph are

$2.00,\ 237,\ 39.5 \times 10^3$, and 98.7×10^3 mm / sec^2 at

frequencies f = 0.1, 1.0, 10, and 50 Hz, respectively.

The largest magnitude is A = 98.7 m/sec at f = 50 Hz. Note that the linear scale indicates a zero displacement amplitude at 100 Hz, however, even a very small displacement amplitude at higher frequencies can contribute significant acceleration.

9.15 **d)** The stiffness required is the smallest value consistent with the maximum displacement constraint,

$$k = F / x_{max} = (mg) / x_{max} = (1 \times 9.80) / \left(5.0 \times 10^{-3}\right) = 1.96 \times 10^{3} \text{ N / m}$$

9.16 **c)** The natural frequency (Eq.9.1.5) of the isolated engine controller is

$$\omega_n = \sqrt{k/m} = \sqrt{(2 \text{ N/mm})(1{,}000 \text{ mm/m})/(1 \text{ kg})}$$
$$= 44.7 \text{ rad/sec} = 7.12 \text{ cyc/sec}$$

9.17 **d)** The isolator transmissibility T is found using Eq.9.6.1 with $\omega_n = 44.7$ rad/sec = 7.12 Hz. The frequency ratio in this case is

$$f/f_n = 50/7.12 = 7.02$$

9.18 **c)** The relationship, $(\omega/\omega_n) > \sqrt{2}$ yields
$\omega_n^2 = \left[(50 \text{ cyc/sec}) \times (2\pi \text{ rad/cyc })\right]^2/2 = 49.3 \times 10^3 \text{ rad}^2/\text{sec}^2$ and
solving for the stiffness using (9.1.5) gives
$k = \omega_n^2 m = \left(49.3 \times 10^3 \text{ rad}^2/\text{sec}^2\right)(1 \text{ kg}) = 49.3 \times 10^3 \text{ N/m} = 49.3 \text{ N/mm}$

9.19 **a)** The relationship (Eq.9.1.5)for Transmissibility T shows that transmissibility is reduced by increasing frequency ratio(ω/ω_n) and decreasing ζ for $(\omega/\omega_n) > \sqrt{2}$. The frequency ratio is increased by decreasing stiffness k to decrease ω_n. The answer is minimize both stiffness and damping ratio ζ.

Machine Design

by Clark J. Radcliffe and Charles W. Radcliffe

Solutions to Practice Problems

Essay Problem Solutions

10.1

A preload of 90% of the bolt's Proof Strength S_P will be applied as a first trial. For a solution, the bolt design must be evaluated iteratively. Assume a bolt length of 1.5 inches long to allow for the nut. First use Tables 10.1 and 10.3 and select a trial design of a 1/2 inch diameter UNC fastener of SAE grade 1. Table 10.3 shows the Proof Strength to be 33 kpsi. The tensile area from Table 10.1 is 0.1419 in^2. The preload is

$$F_i = 0.9 S_P A_t = 0.9 \times 33{,}000 \times 0.1419 = 4214 \text{ lb}$$

The lengths of the unthreaded shank and exposed thread are computed using Figure 10.3. The unthreaded shank length

$$L_u = L_{Total} - L_{thread} = 1.5 - [2(0.5) + 0.25] = 0.25 \text{ in}$$

while the exposed thread length

$$L_t^* = L_{\text{grip}} - L_u = 1.0 - 0.25 = 0.75 \text{ in}$$

Using Eq.10.1.2, the bolt stiffness

$$k_B = \left[\frac{1}{k_u}+\frac{1}{k_t}\right]^{-1} = \left[\frac{L_u}{A_u E}+\frac{L_t^*}{A_t E}\right]^{-1} = \left[\frac{0.25}{(0.1963)(30\times10^6)}+\frac{0.75}{(0.1419)(30\times10^6)}\right]^{-1} = 4.57\times10^6$$

Using Eq.10.1.3, the grip member stiffness

$$k_m = \frac{0.577\pi Ed}{2\ln\left(5\dfrac{0.577L_{\text{grip}}+0.5d}{0.577L_{\text{grip}}+2.5d}\right)} = \frac{0.577\pi(30\times10^6)(0.5)}{2\ln\left(5\dfrac{0.577(1.0)+0.5(0.5)}{0.577(1.0)+2.5(0.5)}\right)} = 16.6\times10^6$$

The joint stiffness factor

$$C = \frac{k_B}{k_B+k_m} = \frac{4.57\times10^6}{4.57\times10^6+16.6\times10^6} = 0.216$$

allows the computation of the load in the bolt (Eq.10.1.4) and member (Eq.10.1.5) under the load $P = 1000$ lb.

$$F_b = \frac{k_B}{k_B+k_m}P + F_i = 0.216(1,000)+4,214 = 4,430 \text{ lb and}$$

$$F_m = \frac{k_m}{k_b+k_m}P - F_i = (1-0.216)(1,000)-4214 = -3430 \text{ lb.}$$

Using the 1/2 - UNC Tensile Stress Area in Table 10.1 and Bolt Strength Standards in Table 10.3, the maximum tensile stress on the bolt and yield factor-of-safety are

$$\sigma = \frac{F_b}{A_t} = \frac{4,430}{0.1419} = 31.2 \text{ kpsi and } n_{Yield} = \frac{S_y}{\sigma} = \frac{36}{31.2} = 1.15$$

The load required to separate the joint and the joint separation factor-of-safety are

$$P_0 = \frac{F_i}{1-C} = \frac{4,214}{1-0.216} = 5,375 \text{ lb. and } n_{sep} = \frac{P_0}{P} = \frac{5,375}{1,000} = 5.38$$

These two factors-of-safety are quite different. To increase the yield factor of safety, decrease the initial preload. Decreased preload will reduce the joint separation factor of safety. Repeated trials to balance the factors of safety indicate a preload of about 40% of Proof Strength. The preload is then

$$F_i = 0.65S_P A_t = 0.40\times33,000\times0.1419 = 1873 \text{ lb}$$

with $\sigma = \dfrac{F_b}{A_t} = \dfrac{2,089}{0.1419} = 14.7$ kpsi and $n_{Yield} = \dfrac{S_Y}{\sigma} = \dfrac{36}{14.7} = 2.45.$

The load required to separate the joint and the joint separation factor of safety are

$$P_0 = \frac{F_i}{1-C} = \frac{1,873}{1-0.216} = 2,389 \text{ lb. and } n_{sep} = \frac{P_0}{P} = \frac{2,389}{1,000} = 2.39$$

Using Table 10.5 for a Cadmium plated bolt and nut, the required preload torque (Eq.10.1.7) for this bolted joint is

$$T = KF_i d = 0.16(2,389)0.5 = 191 \text{ in - lb}$$

and the appropriate SAE bolts specification for machine tolerances is a

1/2 - 13 UNC - 2A Grade 1 bolt, 1.5 ″ long, preloaded with a 191 in-lb torque.

10.2

Natural frequency of the car/spring

$$\omega = 2Hz = 4\pi \text{ (rad / sec)} = 12.57 \text{ (rad/ sec)} = \sqrt{2K/M}$$

The factor of 2 arises from the two springs acting in parallel on the vehicle. The vehicle mass $M = 1200 / 32.2 = 37.27$ slugs so that the required stiffness is

$$K = \omega^2 M/2 = (12.57)^2 (37.27)/2 = 2.94 \times 10^3 \text{ lb / ft} = 245 \text{ lb / in}$$

a) Using Eq.10.2.2

$$\tau_{max} = K_s \frac{8FD}{\pi d^3} \text{ where } F = (1200 / 2) + (1.5)(245) = 968 \text{ lb},\ D = 4.5 \text{ in},$$

d is unknown, and using Eq.10.2.4 $K_s = (2C+1)/2C$ where the spring index $C = D / d$. Setting τ_{max} = 95,000 and solving iteratively (trial and error) for d yields $d = 0.498$ in.

The closest standard size is $d = 0.500$ in. which reduces the stress to 93.7 kpsi.

b) Rearranging the helical compression spring stiffness given by Eq.10.2.8 and solving for active turns,

$$N_a = \frac{(0.5 \text{ in})^4 (11.5 \times 10^6 \text{psi})}{8(4.5 \text{ in})^3 (245.2 \text{ lb/in})} = 4.02 \text{ active turns}$$

For square ends, add 2 additional turns (Eq.10.2.9) yielding $N = 6$ total turns on the springs with square ends.

c) The minimum free length is the static deflection plus maximum deflection plus the summation of the wire coils,

$$L_o = \text{static} + \text{dynamic} + \text{wire} = 600/245.2 + 1.5 + 6 \times 0.5 = 6.96 \text{ in}$$

10.3

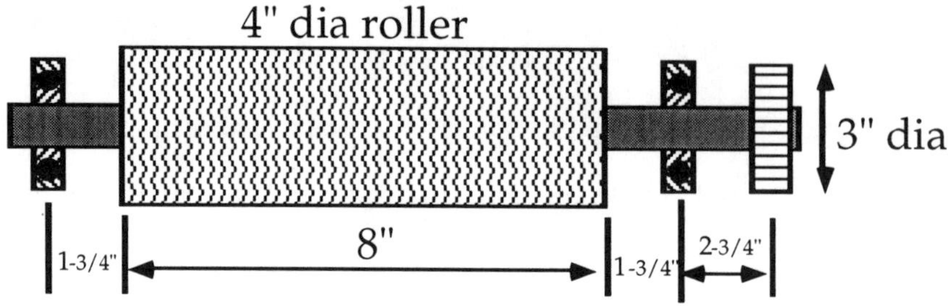

The applied load on the roller is a combination of normal load $F_n = (8 \text{ in})(30 \text{ lb/in}) = 240 \text{ lb}$ given by the roller load/unit length and tangential load $F_t = mF_n = (0.4)240 \text{ lb} = 96 \text{ lb}$ limited by the coefficient of friction. The tangential load generates a torque on the roller which is supplied by a second tangential load on the drive gear, $F_{gear} = 96 \text{ lb}(4 \text{ in}/3 \text{ in}) = 128 \text{ lb}$. The orientation of the drive gear is unspecified so assume a "worst case" and orient that force parallel to the largest roller force to draw the free-body diagram below.

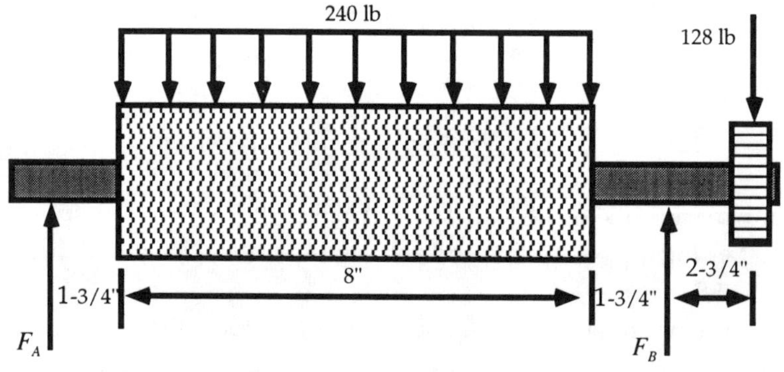

The bearing reaction forces are found by summing moments about "B" to find

$$F_A = [(240 \text{ lb})(4" + 1.75") - (128 \text{ lb})(2.75")]/(1.75" + 8" + 1.75") = 89.4 \text{ lb}$$

and by summing moments about "A" to find

$$F_B = [(240 \text{ lb})(4" + 1.75") + (128 \text{ lb})(1.75" + 8" + 1.75" + 2.75")]/(1.75" + 8" + 1.75") = 278.6 \text{ lb}$$

Shaft Bending and Shear Stresses (see Sect 8.1) are computed for the shaft in the middle of the roller using a free-body diagram for the "sectioned" roller. This location has the largest bending moment.

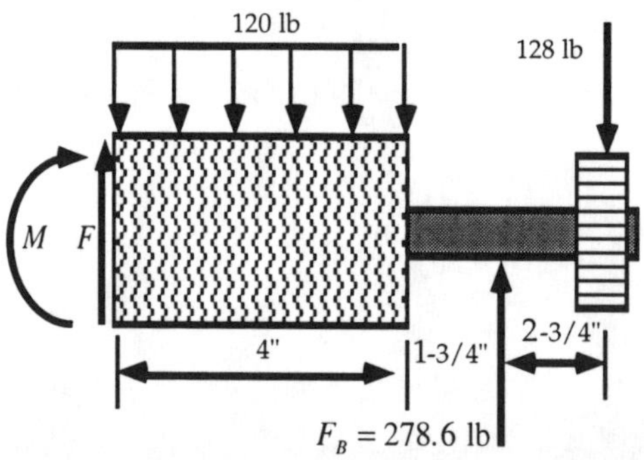

In this case, the shaft shear force $F = 128 \text{ lb} + 120 \text{ lb} - 278.6 \text{ lb} = -30.6 \text{ lb}$ and the shaft bending moment $M = (278.6 \text{ lb})(5.75") - (120 \text{ lb})(2") - (128 \text{ lb})(8.5")$ $= 274 \text{ lb-in}$

These loads result in stresses on the outside surface of the shaft. The bending moment M creates a fully reversed, normal, longitudinal stress σ_a dependent on shaft diameter (see Sect 8.1 , Sect. 10.3 and Equation 10.3.4) while the torsion $T = F_{gear}(d/2) = 192 \text{ in-lb}$ creates a constant normal stress σ_m (Eq.10.3.3) where

$$\sigma_a = 32\frac{M}{\pi d^3} \text{ and } \sigma_m = 27.7\frac{T}{\pi d^3}$$

Using the ASME elliptic failure criteria (Eq.10.3.5) with factor-of-safety $n = 3.5$ yields

$$1 = \left(\frac{n\sigma_a}{S_e}\right)^2 + \left(\frac{n\sigma_m}{S_y}\right)^2 = \left(\frac{32nM}{\pi d^3 S_e}\right)^2 + \left(\frac{27.7nT}{\pi d^3 S_y}\right)^2 \quad \text{or} \quad d^3 = \left(\frac{35.7M}{S_e}\right)^2 + \left(\frac{30.9T}{S_y}\right)^2$$

The yield strength $S_y = 39$ kpsi is provided while the endurance strength S_e of the shaft is found by assuming a constant cross section with a machined finish at room temperature (see Sect 8.4):

$$S_e = 0.5k_ak_bk_ck_dk_eS_u = 0.5(0.869)(1.0)(1.0)(1.0)(1.0)(72 \text{ kpsi}) = 31.3 \text{ kpsi}$$

where the size factor k_b has been neglected. Substituting in the known strengths and loads yields

$$d^3 = \left(\frac{35.7(274 \text{ lb-in})}{31.3 \text{ kpsi}}\right)^2 + \left(\frac{30.9(192 \text{ lb-in})}{39 \text{ kpsi}}\right)^2 = 0.1208 \text{ in}^3 \text{ or } d = 0.494 \text{ in}$$

The shaft should be specified as the next largest standard size so that $d = 1/2$ in

Multiple Choice Problem Solutions

10.4 **d)** Area, $A = \frac{\pi}{4}\left[4.0^2 - 1.5^2\right] = 10.8 \text{ in}^2$ so that

$F_{rod} = 3000 \text{ psi}\left(10.8 \text{ in}^2\right) = 32,400 \text{ lb}$ and

$\sigma_{rod} = (32,400 \text{ lb})\Big/\left(\frac{\pi}{4}1.5^2 \text{ in}\right) = 18,300 \text{ psi}$

10.5 **a)** Because the internal and external pressures are specified in the problem statement, only the cylinder diameter d_{cyl} and thickness t_{wall} are required to compute the "long cylinder" ($l > d_{cyl}$) stresses in the cylinder tube where the stress state is assumed constant over the length of the cylinder.

Note: the "long cylinder" constant stress solution is used for design because it is conservative. The "short cylinder", non-constant stress problem yields smaller stresses because the end caps support the tube, cannot be solved analytically and is not typically used in design.

10.6 **e)** Using a half cylinder free-body diagram and a "thin-wall" constant stress assumption

$$\sigma_{wall} = \frac{F}{A} = \frac{[(d_{cyl}l)P_{internal}]}{[2t_{wall}l]}$$

$$= \frac{[(d_{cyl})P_{internal}]}{[2t_{wall}]} = \frac{[(4.0)3000]}{[2(0.25)]} = 24{,}000 \text{ psi}$$

(The thick-walled, "exact" model uses Lame's equation and yields $\sigma_{wall} = 21{,}300$ psi)

10.7 **a)** Examining the design, the tie rods carry the longitudinal tensile stress due to pressure yielding no longitudinal stress in the cylinder due to pressure. The only stress in the tube is the compressive stress due to any preload of the tie rods BEFORE the tube is pressurized.

10.8 **b)** The axial stress on the tie rods due to the internal pressure and neglecting preload is

$$\sigma_{tie\ rod} = \frac{F}{A} = [32{,}400 / 6] \Big/ \left(\frac{\pi}{4} 0.75^2 \text{ in}\right) = 12{,}200 \text{ psi}$$

10.9 **e)** The shear stress on the piston rod threads is determined by the piston rod diameter d_{rod}, the thread size and length (from h and thread specification), and the piston rod load (from d_{cyl}, d_{rod} and $P_{internal}$). $P_{internal}$ is specified in the problem statement. You need a) d_{cyl}, d) h, e) d_{rod}, and i) the piston thread specification.

10.10 **c)** The thread shear stress acts at half the height of the thread:

$$d_{THD} = 1.0 - \tfrac{1}{8} = 0.875 \text{ in}$$

$$\tau_{THD} = \frac{F}{A} = \frac{F}{[(\pi d_{THD})h/2]} = \frac{2F}{[(\pi d_{THD})h]} = \frac{2(32{,}400)}{[(\pi 0.875)1.5]} = 10{,}400 \text{ psi}$$

10.11 **e)** The rod force is now compressive and due to the internal pressure acting over the full area of the piston,

$$F_{rod} = 3000 \text{ psi}\left(\frac{\pi}{4}(4.0^2) \text{ in}^2\right) = 37{,}700 \text{ lb}$$

10.12 **a)** The tie rod load goes to zero. The hoop stress and longitudinal stress in the cylinder remain unchanged. The shear stress in the threads increases because the compressive longitudinal load is larger than the past tensile load.

Control Systems

by Clark J. Radcliffe and Charles W. Radcliffe

Solutions to Problems

Essay Problem Solutions

11.1 First find the damping ratio ζ associated with the specified peak magnification M_p. The specified peak magnification $M_m = 1.15$ is the maximum magnitude of the closed loop transfer function, $G_{CL} = G_{OL}/(1+G_{OL})$ at some frequency to be determined.

a) For a second order system, the relationship between peak magnification M_p and damping ratio ζ is $M_m = 1\Big/\left[2\zeta\sqrt{1-\zeta^2}\right] \Leftrightarrow \zeta = \sqrt{M_m - \sqrt{M_m^2 - 1}}$. For our case,

$$\zeta = \sqrt{1.155 - \sqrt{1.155^2 - 1}} = \sqrt{1.155 - \sqrt{0.3340}} \cong 0.76$$

b) The open-loop transfer function is $G_{OL} = K_g\big/\left[s(1+0.2s)(1+0.5s)\right]$ where $K_g = C_1C_2$.

Set up the closed-loop transfer function,

$$M_m = \|G_{CL}\| = \left\|\frac{G_{OL}}{1+G_{OL}}\right\| = \left\|\frac{\left[\dfrac{K_g}{s(1+0.2s)(1+0.5s)}\right]}{\left[1+\dfrac{K_g}{s(1+0.2s)(1+0.5s)}\right]}\right\| = \left\|\frac{10K_g}{s^3+7s^2+10s+10K_g}\right\|$$

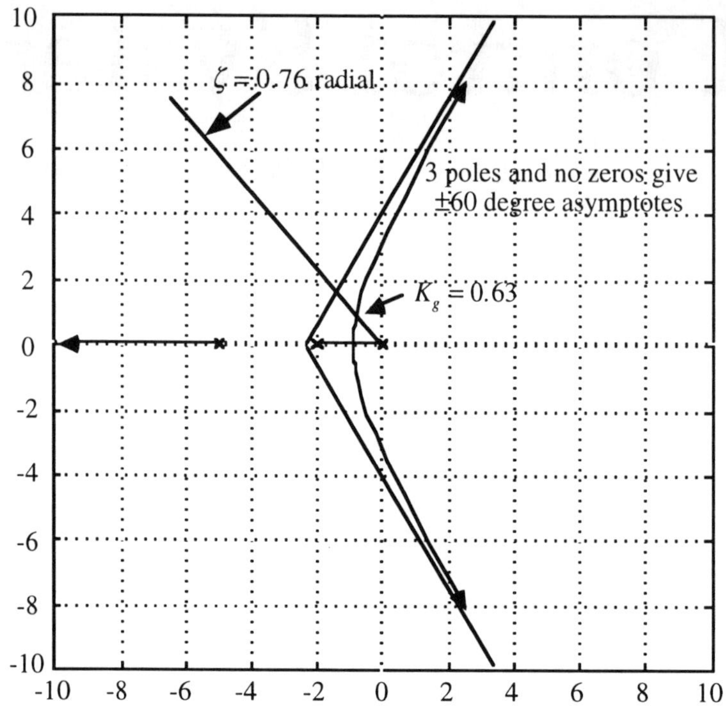

Draw the Root Locus and indicate the radial where pole positions have the required damping ratio, $\zeta = 0.76$ using the special form of the characteristic equation,

$$1 + K_g \frac{10}{s^3 + 7s^2 + 10s} = 1 + K_g \frac{10}{s(s+2)(s+5)} = 0$$

The characteristic equation is third order with one real and two complex conjugate poles.

The Root Locus displays an intersection of the $\zeta = 0.76$ radial at $s \cong 0.82 + 0.70j$. Substituting this value of s into the relationship for control gain yields,

$$K_g = \frac{-s(s+2)(s+5)}{10}$$

$$\cong \frac{-(-0.82 + 0.7i)(-0.82 + 0.7i + 2)(-0.82 + 0.7i + 5)}{10} \cong 0.63$$

Drawing a Bode Diagram yields the gain and phase margins. Using this $K_g = 0.63$ in the open-loop gain, $G_{OL} = 6.3/s(s+2)(s+5)$.

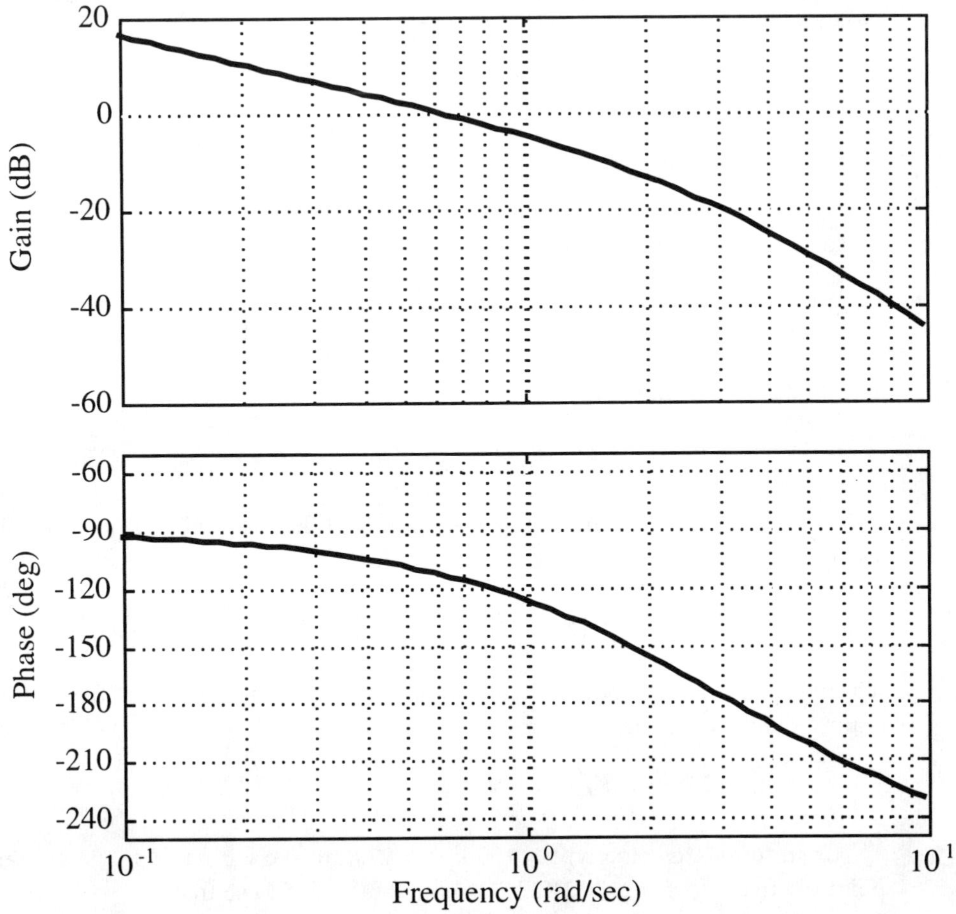

The Bode Diagram shows a gain margin of 20 dB at $\omega = 3$ rad / sec and a phase margin of approximately 70 degrees at $\omega = 0.6$ (rad/sec).

c) The frequency at which the peak closed loop response is expected is at the damped natural frequency associated with the complex conjugate poles for $K_g = 0.63$. Since the closed loop poles are located at $cs = 0.82 + 0.70j$, the damped natural frequency for these poles is at $\omega_m = 0.70$ rad / sec.

11.2

a) The first order, nonlinear, ordinary differential equation system model is

$$m\dot{v} = F_m(t) - F_f(v) - F_a(v) = F_m(t) - K_f v - K_a v^2$$

b) The aerodynamic drag term is the cause of the nonlinearity. Linearization about the "operating point" $v = v_0 =$ 30 meters/sec gives the relationship,

$$F_a(v) = K_a v^2 = F_a(v_0) + \left.\frac{\partial F_a}{\partial v}\right|_{v=v_0} (v - v_0) + HOT$$

$$= K_a v_0^2 + (2K_a v_0)(v - v_0) + HOT$$

A similar expansion for the linear, coulomb friction force has all higher order derivatives in the Taylor expansion equal to zero:

$$F_f(v) = K_f v = F_f(v_0) + \left.\frac{\partial F_a}{\partial v}\right|_{v=v_0}(v - v_0) + HOT$$

$$= K_f v_0 + (K_f)(v - v_0)$$

To complete the linearization, we define a new velocity,

$$\tilde{v} = (v - v_0), \text{ with } \dot{\tilde{v}} = \frac{d(v - v_0)}{dt} = \dot{v}$$

and the ODE becomes

$$m\dot{\tilde{v}} = F_m(t) - (K_f v_0 + K_f \tilde{v}) - (K_a v_0^2 + 2K_a v_0 \tilde{v} + HOT)$$

We now can find an linear approximation to the original ODE by neglecting second order terms.

$$m\dot{\tilde{v}} + (K_f + 2K_a v_0)\tilde{v} = F_m(t) - K_f v_0 - K_a v_0^2 = \tilde{F}_m(t)$$

where $\tilde{F}_m(t) = F_m(t) - K_f v_0 - K_a v_0^2$ is the tractive force deviation from the operating point. We have

$$m = 1000 \text{ kg}, \ K_f = 0.1 \text{ N} \cdot \text{sec} / \text{m}, \text{ and } K_a = 1.0 \text{ N} \cdot \text{sec}^2 / \text{m}^2$$

For steady-state velocity, $v = 100$ kph $= 27.8$ m/sec with $\dot{v} = 0$, the operating point tractive effort F_m=(0.1)(27.8)+(1.0)(27.8^2) = 776 N so that

$$\tilde{F}_m(t) = F_m(t) - 776 \text{ N}$$

For $K_f + 2K_a v_0 = 0.1 + 2(1)(27.8) = 55.7$, the linearized system differential equation becomes

$$1000\dot{\tilde{v}} + 55.7\tilde{v} = \tilde{F}_m(t)$$

which can be expressed in state space form as

$$\dot{\tilde{v}} = -0.0557\tilde{v} + 0.001\tilde{F}_m(t)$$

The system transfer function is found from the Laplace transform of the system model:

$$G(s) = \frac{\tilde{V}(s)}{\tilde{F}_m(s)} = \frac{0.001}{s + 0.0557} = \frac{0.018}{18s + 1}$$

The vehicle model has a single real root at s = -0.0557 and a time constant of 18.0 sec with a steady state gain of 0.018.

c) The open-loop vehicle model is has a single real pole and no poles at the origin so requires integral control action for zero steady-state error. The specification requires a time constant of $\tau = 10$ sec which is faster than the 18 sec open-loop time constant requiring control action proportional to error to speed up the closed loop system response. Propose an Proportional-Integral (PI) control $G_1(s) = K_P + K_I/s$ and sketch the closed-loop block diagram below assuming unity measurement gain.

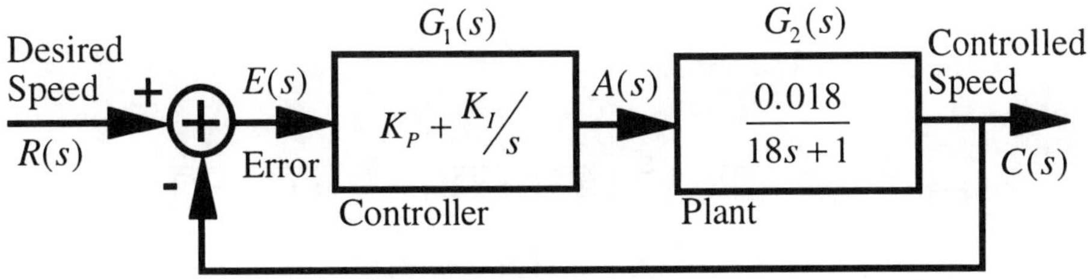

The open-loop transfer function becomes

$$G_{OL}(s) = G_1(s)G_2(s) = K_P\left[\frac{s+a}{s}\right]\left[\frac{0.018}{18s+1}\right] = 0.001K_P\left[\frac{s+a}{s}\right]\left[\frac{1}{s+0.0557}\right]$$

where $a = K_I/K_P$, the ratio of integral to proportional gains. To examine the effect of proportional and integral gains on closed-loop response, sketch the system root locus in terms of the parameter a.

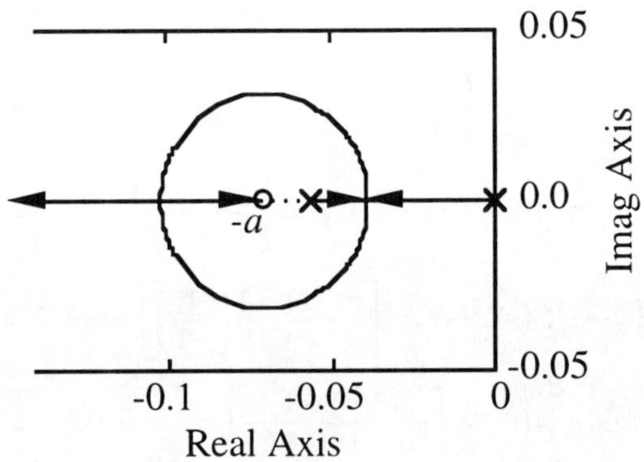

We desire a closed-loop pole at s = -0.1 for $\tau = 10$ sec so place $a = K_I/K_P = 0.07$ then calibrate for K_P where s = -0.1 based on the Root Locus rules. Calibrating the Root Locus (rule 11), for $a = K_I/K_P$ =0.07 at s = -0.1 yields

$$K_P = \left[\frac{-s}{s+a}\right]\left[\frac{18s+1}{0.018}\right] = \left[\frac{0.1}{-0.1+0.07}\right]\left[\frac{18(-0.1)+1}{0.018}\right] = \left[-3.3\bar{3}1\right]\left[-44.\bar{4}\right] = 148$$

and for $a = K_I/K_P$ =0.07

$$K_I = 0.07K_P = 10.4$$

For the system transfer function,

$$G(s) = \frac{Y(s)}{F(s)} = \frac{3s+1}{(s+1)(s+2)}$$

Multiple Choice Problem Solutions

11.3 **b)** The impulse response is

$$g(t) = \mathcal{L}^{-1}(G(s)) = \mathcal{L}^{-1}\left(\frac{3s+1}{(s+1)(s+2)}\right) = \mathcal{L}^{-1}\left(\frac{k_1}{s+1} + \frac{k_2}{s+2}\right)$$

where $k_1 = \lim_{s\to -1}((s+1)G(s)) = \left[\frac{3(-1)+1}{(-1+2)}\right] = -2$

$$k_2 = \lim_{s\to -2}((s+2)G(s)) = \left[\frac{3(-2)+1}{(-2+1)}\right] = 5$$

so that

$$g(t) = -2e^{-t} + 5e^{-2t}$$

11.4 **c)** The step response is

$$y(t) = \mathcal{L}^{-1}(G(s)U(s)) = \mathcal{L}^{-1}\left(\frac{3s+1}{(s+1)(s+2)}\frac{1}{s}\right) = \mathcal{L}^{-1}\left(\frac{k_1}{s+1} + \frac{k_2}{s+2} + \frac{k_3}{s}\right)$$

where $k_1 = \lim_{s\to -1}((s+1)G(s)U(s)) = \left[\frac{3(-1)+1}{(-1+2)}\left(\frac{1}{-1}\right)\right] = 2$

$$k_2 = \lim_{s\to -2}((s+2)G(s)U(s)) = \left[\frac{3(-2)+1}{(-2+1)}\left(\frac{1}{-2}\right)\right] = -5/2 = -2.5$$

$$k_3 = \lim_{s\to 0}((s)G(s)U(s)) = \left[\frac{3(0)+1}{(0+1)(0+2)}\right] = 1/2 = 0.5$$

so that

$$y(t) = 2e^{-t} - 2.5e^{-2t} + 0.5$$

11.5 **a)** The system's steady state gain is

$$G(s=0) = (3*0+1)/((0+1)(0+2)) = 1/(1*2) = 1/2 = 0.5$$

11.6 **a)** The frequency response of the system $Y(\omega)/F(\omega)$ at $\omega = 2\pi f = 2\pi(3) = 18.85$ rad / sec is

$$\frac{Y(\omega)}{F(\omega)} = \frac{(3)18.85j+1}{(18.85j+1)(18.85j+2)} = 0.0222 - 0.1565j = 0.1581e^{-1.4298j}$$

The frequency response has a magnitude ratio, $\|Y(\omega)/F(\omega)\| = 0.1581$ and a phase angle,

$$\text{angle}[Y(\omega)/F(\omega)] = -1.4298 \text{ rad} = -81.92 \text{ deg}$$

11.7 **b)** The system's order is 2 based on the order of the denominator polynomial, choice.

11.8 **a)** The system's two poles are both "negative real" so the system is "stable".

11.9 **c)** The (2) poles are at $s = -1$ and -2 .

11.10 **d)** The single zero is at $s = -1/3$.

11.11 **a)** To determine closed-loop stability, sketch the root-locus:

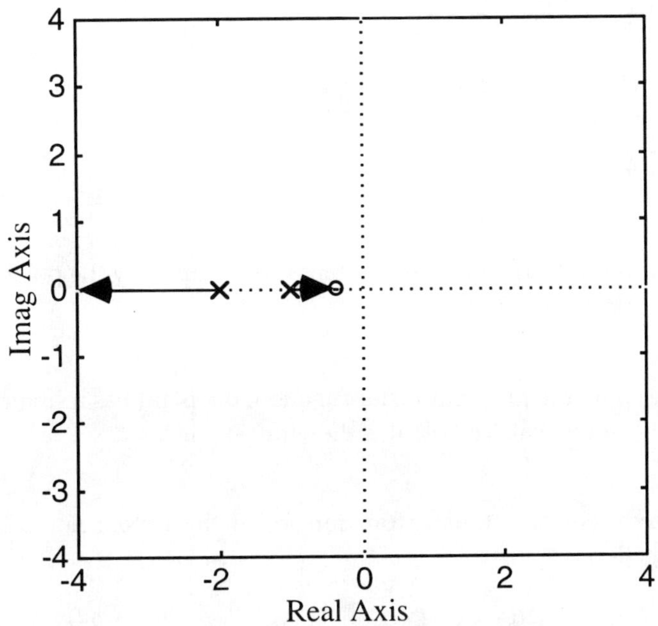

The system is closed loop stable for all positive values of k_p. To examine the stability for negative values of k_p, write the closed-loop characteristic equation, $1 + k_p G_c G_p H = 0$ as

$$1 + k_p \frac{3s+1}{(s+1)(s+2)} = 0 \Rightarrow (s+1)(s+2) + k_p(3s+1) = 0 \quad \text{or}$$

$$s^2 + \left(3 + 3k_p\right)s + \left(2 + k_p\right) = 0$$

which has negative real roots when

$$\left(3 + 3k_p\right) > 0 \Rightarrow k_p > -1 \text{ and } \left(2 + k_p\right) > 0 \Rightarrow k_p > -2$$

Simultaneous application of the two conditions requires

$$k_p > -1$$

11.12 **e)** The root locus and stability analysis above indicates the system's least stable positive control gain is $K_p = \infty$.

11.13 **b)** For the plant model differential equation given, the Laplace transform is $sC(s) = -3C(s) + 5A(s)$ yielding the plant transfer function

$C(s)/A(s) = 5/(s+3)$

11.14 **d)** The steady-state gain of the plant is the value of the plant transfer function at $s = 0$,

$C(0)/A(0) = 5/(0+3) = 5/3$

11.15 **c)** The frequency response at 4 rad/sec is found by evaluating the plant transfer function for $s = 4j$:

$$\frac{C(s)}{A(s)} = \frac{C(4j)}{A(4j)} = \frac{5}{(4j+3)}$$

which has magnitude,

$$\left\| \frac{C(4j)}{A(4j)} \right\| = \frac{5}{\sqrt{4^2+3^2}} = 1$$

11.16 **b)** The plant's slowest time constant is developed by the pole at $s = -3$, $\tau = 1/3 \text{ sec} = 0.33 \text{ sec}$.

11.17 **a)** Based on the plant's transfer function, the plant has a single pole which is negative real. The plant is therefore stable.

11.18 **a)** Using the plant transfer function above, the system closed-loop transfer function is

$$T(s) = \frac{C(s)}{R(s)} = \frac{KG}{1+KGH} = \left[\frac{5K}{(s+3)}\right] \Big/ \left[1 + \frac{25K}{(s+3)(s+5)}\right]$$

which simplifies to

$$T(s) = \frac{5K(s+5)}{(s^2+8s+15)+25K}$$

11.19 **a)** The closed-loop system steady-state gain is the value of the closed-loop transfer function at $s = 0$,

$T(0) = 5K(0+5) \Big/ \left[(0^2 + 8 \times 0 + 15) + 25K\right] = 25K/[15+25K]$,

11.20 **c)** The closed-loop frequency response at 4 rad/sec is found by evaluating the closed-loop transfer function for $s = 4j$,

$$T(4j) = \frac{5K(4j+5)}{\left[(-16+8(4j)+15)+25K\right]} = \frac{(25+20j)}{[24+32j]}$$

which has magnitude,

$$\|T(4j)\| = \sqrt{(25^2+20^2)}\Big/\sqrt{24^2+32^2} = 1025/1600 = 0.64$$

11.21 **c)** The closed-loop system's slowest time constant can be identified through a root locus. First the closed-loop characteristic equation is written in the standard form as $1 + KGH = 0 \Rightarrow 1 + \frac{25K}{(s+3)(s+5)} = 0$ where $N(s) = 25$ and $D(s) = (s+3)(s+5) = s^2 + 8s + 15$ The sketch below shows the resulting Root Locus.

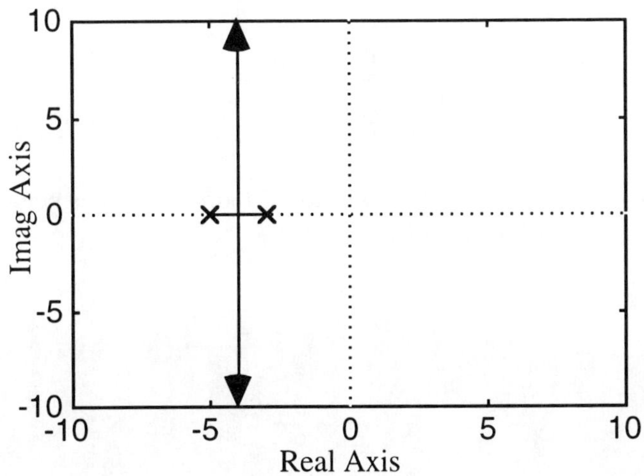

To calibrate the Root Locus, select values of the Laplace variable "s" and compute the Gain, $K = -D(s)/N(s)$ on the indicated paths of the Root Loci. At $K = 1$, both closed-loop poles are equidistant from the imaginary axis at $s \cong -4 \pm 5j$ and contribute equally slow components to the time response.

s	K
$-3.0 + 0.0j$	0
$-3.5 + 0.0j$	0.03
$-4.0 + 0.0j$	0.04
$-4.0 + 0.5j$	0.05
$-4.0 + 1.0j$	0.08
$-4.0 + 1.5j$	0.13
$-4.0 + 2.0j$	0.20
$-4.0 + 2.5j$	0.29
$-4.0 + 3.0j$	0.40
$-4.0 + 3.5j$	0.53
$-4.0 + 4.0j$	0.68
$-4.0 + 4.5j$	0.85
$-4.0 + 5.0j$	1.04

11.22 **c)** The Root Locus indicates that the closed-loop system has negative-real poles for all positive gains K.

Heat Transfer

by Keith Woodbury

Solutions to Practice Problems

Essay Problem Solutions

12.1 a) $q_{overall} = q_{surface}$ $\qquad \dfrac{36-(-5)}{\dfrac{0.003}{0.2}+\dfrac{1}{10}} = \dfrac{T_s-(-5)}{\dfrac{1}{10}} \qquad \Rightarrow T_s = 30.7\text{ C}$

b) $T_s = 18.4\text{ C}$

c) $q_{windy} = \dfrac{36-T_\infty}{\left(\sum R_{th}\right)_{calm}} \qquad \dfrac{36-(-5)}{\dfrac{0.003}{0.2}+\dfrac{1}{50}} = \dfrac{36-T_\infty}{\dfrac{0.003}{0.2}+\dfrac{1}{10}} \qquad \Rightarrow T_\infty = -99\text{ C}$

12.2 a) $P_i = hA(T_s - T_\infty) \qquad T_s = 0.75/(20\times 0.03\times 0.03)+30 = 72\text{ C}$

b) Different assumptions about the tip condition are possible. Assume no heat loss from the tip (insulated tip). Then the heat dissipated by one fin follows from the first case in Table 12.5:

$q_{fin} = M\tanh(mL)$

$$M = \sqrt{20\cdot(0.004\pi)\cdot 177\cdot(\pi 0.004^2/4)}(50-30) = 0.4729$$

$$m = \sqrt{20\cdot(0.004\pi)/\left[177\cdot(\pi 0.004^2/4)\right]} = 10.63$$

$q_{fin} = 0.4729\times\tanh(10.63\times 0.015) = 0.0748\text{ W}$

$q_{total} = q_{fins} + q_{no\,fins} = N\cdot q_{fin} + (w^2 - N\cdot\pi d^2/4)h\Delta T = 0.75 \qquad \Rightarrow N = 5.6 \approx 6$

12.3 a) All of the heat generated must be conducted through the steel:

$$q'''A\cdot L = \frac{T_{max} - T_\infty}{\dfrac{\ln((d_i + 2\Delta r)/d_i)}{2\pi kL} + \dfrac{1}{h\pi d_o L}} = \frac{T_{max} - 90}{\dfrac{\ln(17/15)}{2\pi(14.9)L} + \dfrac{1}{5000\cdot\pi\cdot(0.017)L}}$$

$\Rightarrow T_{max,st} = 719\text{ C}$

b) Maximum temperature occurs at centerline. Use equation for solid cylinder with constant internal heat generation:

$$T(0) = T_{max,st} + \frac{700\times 6\cdot(0.015^2)}{4\cdot 43.9} = 1616\text{ C}$$

12.4 a) Assume the average surface temperature is 300 C, then the average film temperature is $T_{film} = (600{+}300)/2 = 450$ C. Then the air properties are:

$\nu = 71.9\times 10^{-6}\ \text{m}^2/\text{s} \qquad k = 0.0536\ \text{W/m}\cdot\text{C} \qquad \text{Pr} = 0.698$

Now compute Reynolds number to determine the flow regime:

$$\text{Re}_L = \frac{V\cdot L}{\nu} = \frac{10\times 1.50}{71.9\times 10^{-6}} = 208\,600 < 500\,000 \quad \text{so laminar over entire length}$$

Then

$$\overline{\text{Nu}}_L = 0.664\cdot 208\,600^{0.5}\cdot 0.698^{1/3} = 269 \quad \Rightarrow \bar{h} = 269\cdot 0.0536/1.50 = 9.6\ \text{W/m}^2\cdot\text{C}$$

b) The center of the slab will take the longest to heat. Assume all faces of slab convect heat with same h just computed. Check for lumped capacitance assumption:

$$L_c = V/A = (1.50\cdot 5.0\cdot 0.6)/2/(1.5\cdot 5 + 1.5\cdot 0.6 + 5.0\cdot 0.6) = 0.197\text{ m}$$

$$\text{Bi} = \frac{h(V/A)}{k_{Al}} = \frac{9.6\cdot(0.197)}{177} = 0.01 < 0.1 \text{ Lumped Capacitance O.K.}$$

$$t = \frac{-\rho c_p L_c}{h}\ln\left(\frac{\theta(t)}{\theta_0}\right) = \frac{-2770\cdot 875\cdot 0.197}{9.6}\ln\left(\frac{500-600}{30-600}\right) = 86{,}600\text{ s} = 24\text{ h}$$

12.5 a) The heat transfer coefficient on each side must be found from appropriate correlations. Assume the average wall temperature is about 100 C.

Then the approximate film temperature on the inside of the tube is (600 + 100)/2 = 350 C. Properties of air are

$\nu = 56.15\times10^{-6}\ \mathrm{m^2/s}$ $\quad k = 0.0482\ \mathrm{W/m\cdot C}$ $\quad \mathrm{Pr} = 0.687$

$\rho = 0.5598\ \mathrm{kg/m^3}$ $\quad c_p = 1057\ \mathrm{J/kg\cdot C}$

Then

$$\mathrm{Re}_D = \frac{VD}{\nu} = \frac{\dot{m}}{\rho\pi D^2/4}\cdot\frac{D}{\nu} = \frac{4\cdot 0.5}{0.5598\cdot\pi\cdot 0.5\cdot 56.15\times10^{-6}} = 40\,500 > 2300$$

$$Nu_D = 0.023\times 40\,500^{0.8}\times 0.687^{1/3} = 98.5 \quad \Rightarrow h = \frac{98.5\cdot 0.0482}{0.5} = 9.5\ \mathrm{W/m^2\cdot C}$$

For the outside of the cylinder, the film temperature is (5 + 100)/2 ≈ 50 C.

$\nu = 18.20\times10^{-6}\ \mathrm{m^2/s}$ $\quad k = 0.0280\ \mathrm{W/m\cdot C}$ $\quad \mathrm{Pr} = 0.704$

Then

$$\mathrm{Re}_D = \frac{VD}{\nu} = \frac{5\cdot 0.5}{18.2\times10^{-6}} = 137\,360$$

From Table 12.8, $C = 0.027$ and $m = 0.805$, and

$$\mathrm{Nu}_D = 0.027\cdot 137\,360^{0.805}\times 0.704^{1/3} = 329 \quad \Rightarrow h = \frac{329\cdot 0.0280}{0.5} = 18.4\ \mathrm{W/m^2\cdot C}$$

Now, the overall heat transfer coefficient is

$$UA = \frac{1}{\sum R_{th}} = \left[\frac{1}{h_i}+\frac{1}{h_o}\right]^{-1} A = \left[\frac{1}{9.5}+\frac{1}{18.4}\right]^{-1} A = 6.3A\ \ \mathrm{(W/C)}$$

And

$$\dot{m}c_p(T_{out}-T_{in}) = -UA\frac{((T_{out}-T_\infty)-(T_{in}-T_\infty))}{\ln\left(\dfrac{T_{out}-T_\infty}{T_{in}-T_\infty}\right)}$$

$$\Rightarrow T_{out} = (T_{in}-T_\infty)\times\exp\left[-\frac{UA}{\dot{m}c_p}\right]+T_\infty = 536\ \mathrm{C}$$

b) At the exit:

$$\frac{(T_{out}-T_\infty)}{A\sum R_{th}} = (T_{wall}-T_\infty)h_o A \quad \Rightarrow T_{wall} = \frac{U}{h_o}(536-5)+5 = 187\ \mathrm{C}$$

12.6 a) The steam will maintain constant temperature as long as it undergoes phase change. The temperature will be the saturation temperature at the given pressure. From steam tables, $T_{steam} \approx 106.9$ C. Since the steam temperature is constant, $C_r = 0$, and the water is the minimum fluid. For the exchanger, the effectiveness is

$$\varepsilon = \frac{q}{q_{max}} = \frac{\dot{C}_{water}(T_{water,out}-T_{water,in})}{\dot{C}_{min}(T_{steam,in}-T_{water,in})} = \frac{(60-20)}{(106.9-20)} = 0.460$$

When one fluid temperature is constant,

$$NTU = -\ln(1-\varepsilon) = -\ln(1-.460) = 0.616 = UA/\dot{C}_{\min}$$

$$\Rightarrow A = NTU \cdot \dot{C}_{\min}/U = 0.616 \cdot \left(\frac{8000}{3600}4179\right)/2000 = 2.86\ \text{m}^2$$

b) $\dot{m}_{steam} h_{fg} = \dot{C}_{water}(T_{out} - T_{in})$ From steam tables $h_{fg} \approx 2238$ kJ/kg

$$\Rightarrow \dot{m}_{steam} = \frac{1}{2238}\frac{8000}{3600} \cdot 4.179(60-20) = 0.165\ \text{kg/s} = 600\ \text{kg/hr}$$

12.7 a) The heat transfer coefficient on the inside of the pipe must be found. Assume the pipe wall temperature is about 90 C, then the average film temperature of the oil is 100 C. For unused engine oil:

$\nu = 20.5 \times 10^{-6}\ \text{m}^2/\text{s}$ $\quad k = 0.1367\ \text{W/m}\cdot\text{C}$ $\quad \Pr = 280$

$\rho = 840.1\ \text{kg/m}^3$ $\quad c_p = 2219\ \text{J/kg}\cdot\text{C}$

Then the Reynolds number is

$$\text{Re}_D = \frac{VD}{\nu} = \frac{\dot{m}D}{\rho\pi\left(D^2/4\right)\nu} = \frac{4 \cdot 1.5}{840.1 \cdot \pi \cdot 0.030 \cdot 20.5 \times 10^{-6}} = 3696 > 2300$$

And the Nusselt number is

$$\text{Nu}_D = 0.023 \times 3700^{0.8} \times 280^{1/3} = 108$$

$$\Rightarrow h = 108 \times 0.1367/0.03 = 491\ \text{W/m}^2\cdot\text{C}$$

Then the overall heat transfer coefficient is

$$UA = \left[\frac{1}{\pi \cdot 0.03 \cdot L \cdot 491} + \frac{\ln(36/30)}{2\pi \cdot 60 \cdot L} + \frac{1}{\pi \cdot 0.036 \cdot L \cdot 200}\right]^{-1}$$

$$= L[0.0216 + 0.00048 + 0.442]^{-1} = 15.1L$$

The heat exchanger factor F is unity since one fluid temperature is constant, so

$$L = \dot{m}c_p \Delta T_{oil} = \Delta T_{LM}(UA/L)L$$

$$\Delta T_{LM} = (200-120)/\ln((200-27)/(120-27)) = 128.9$$

$$L = 1.5 \times 2219 \times 80/128.9/15.1 = 137\ \text{m}$$

b) Need the efficiency for one of the fins. Use Figure 12.1:

$$L_c^{3/2}(h/kA_p)^{1/2} = (0.035 + 0.003/5)^{3/2}(200/(60 \cdot 0.00011))^{1/2} = 1.217$$

$$r_{2c}/r_1 = (2 \cdot (3.5 + 0.3) + 3.0)/3.6 = 2.94$$

Entering Figure 12.1, find h_{fin}= 0.44. Now find the resistance of the bank of fins:

$$A_{fin} = 2 \cdot \pi\left(D_o^2 - d_o^2\right)/4 + \pi D_o t = 166\ \text{cm}^2$$

$$A_{space} = \pi d_o s = 5.7\ \text{cm}^2$$

$$\frac{A_{fin}}{A_{total}} = \frac{A_{fin}}{A_{fin} + A_{space}} = 0.967$$

$$\eta_{total} = 1 - \frac{A_{fin}}{A_{total}}(1 - \eta_{fin}) = 0.459$$

Now the overall fin bank resistance is found. The key is to express the total area for heat transfer per unit length:

$A_{total} / L = (A_{fin} + A_{space})/(s+t) = 2.15 \text{ m}^2/\text{m}$

$R_{bank} = L[h(A_{total} / L)\eta_{total}]^{-1} = 0.00507L \text{ C}\cdot\text{m}^2/\text{W}$

Assuming the interior heat transfer coefficient is unchanged, and neglecting the thermal resistance of the pipe wall itself, the overall heat transfer coefficient per unit length is

$UA / L = [1/\pi d_i h_i + R_{bank} / L]^{-1} = 37.5$

And

$L = L_{no\,fins} \times (15.1/37.5) = 44 \text{ m}$

12.8 a) Need heat transfer coefficients for both sides of the tubes. Assume the tube wall temperature is about 100 C, then the average film temperature of the water could be about 75 C:

$\nu = 0.3842 \times 10^{-6} \text{ m}^2/\text{s} \qquad k = 0.668 \text{ W/m}\cdot\text{C} \qquad \text{Pr} = 2.35$

$\rho = 974.9 \text{ kg/m}^3 \qquad c_p = 4193 \text{ J/kg}\cdot\text{C}$

Reynolds and Nusselt numbers:

$$\text{Re}_D = \frac{4\dot{m}}{\rho\pi D\nu} = \frac{4(0.025)}{974.9\cdot\pi\cdot 0.025\cdot 0.3842\times 10^{-6}} = 3400 > 2300$$

$$\text{Nu}_D = 0.023\cdot 3400^{0.8} 2.35^{1/3} = 20.4$$

$$\Rightarrow h_i = 20.4\cdot 0.668/0.025 = 545 \text{ W/m}^2\cdot\text{C}$$

For the air on the exterior of the tubes, the film temperature may be 300 C:

$\nu = 48.85 \times 10^{-6} \text{ m}^2/\text{s} \qquad k = 0.0453 \text{ W/m}\cdot\text{C} \qquad \text{Pr} = 0.684$

$\rho = 0.6088 \text{ kg/m}^3 \qquad c_p = 1045 \text{ J/kg}\cdot\text{C}$

Since the bank is an aligned arrangement, the maximum Re occurs at the transverse section:

$$\text{Re}_{\max} = \frac{S_T}{S_T - D}\frac{V_\infty D}{\nu} = \frac{50}{50-25}\frac{5\cdot 0.025}{48.85\times 10^{-6}} = 5125$$

In Table 12.9, find $C = 0.229$ and $m = 0.632$ (since $S_T/D = S_L/D = 2.0$). Then

$$\overline{\text{Nu}}_D = 1.13 \times 0.229 \times 5{,}125^{0.632} \times 0.684^{1/3} = 50.4$$

$$\Rightarrow \bar{h}_o = 50.4 \times 0.0453/0.025 = 91.3 \text{ W/m}^2\cdot\text{C}$$

Then

$$UA = \left[1/h_i A + 1/h_o A\right]^{-1} = 78.2A = 78.2\cdot N_{tubes}\pi DL = 2460 \text{ W/C}$$

b) Find the effectiveness from the NTU:

$\dot{C}_{air} = 2.25\cdot 1045 = 2350 \text{ W/C} \qquad \dot{C}_{water} = 100\cdot 0.025\cdot 4193 = 10\,500$

So the air is the minimum fluid and $C_r = 0.224$. Also

$$NTU = \frac{UA}{\dot{C}_{\min}} = \frac{2460}{2350} = 1.05$$

For a cross flow exchanger, with one fluid mixed (air) and one unmixed (water), the effectiveness $\varepsilon = 0.6$. So

$$\Delta T_{air} = \frac{\varepsilon \dot{C}_{\min} \Delta T_{\max}}{\dot{C}_{air}} = 0.6 \times (550 - 25) = 315 \text{ C} \quad \Rightarrow \; T_{air,out} = 235 \text{ C}$$

$$\Delta T_{water} = \frac{\varepsilon \dot{C}_{\min} \Delta T_{\max}}{\dot{C}_{water}} = 0.6 \times 0.224 \times (550 - 25) = 71 \text{ C} \quad \Rightarrow T_{water,out} = 96 \text{ C}$$

12.9 a) Key is recognition that the view factor $F_{1\text{-}2} = 90/360{=}0.25$ (25% of the energy leaving the heater rod is intercepted by the floor). The analogous network for this problem is the same as that in Figure 12.16. The equivalent resistance of the spatial resistors is:

$$\frac{1}{R_{eq}} = \left[A_1 F_{1-2} + \frac{1}{1/A_1 F_{1-3} + 1/A_2 F_{2-3}} \right]$$

$$A_1 F_{1-2} = A_2 F_{2-1}; \; F_{2\text{-}3} = 1 - F_{2-1} = 1 - 0.00785 = 0.992$$

$$\frac{1}{R_{eq}} = \left[0.00785 + \frac{1}{42.4 + 1.01} \right] = 0.0309 \quad \Rightarrow R_{eq} = 32.4$$

Then

$$\frac{\dot{P}}{L} = \frac{q_2}{L} = -\frac{q_1}{L} = \frac{E_{b_1} - E_{b_2}}{R_{eq} + (1-\varepsilon_2)/\varepsilon_2 A_2} = \frac{5.67E{-}8(1473^4 - 573^4)}{32.4 + 0.4/(0.6 \cdot (1))} = 7.89 \text{ kW/m}$$

b) The radiosities must be found:

$$J_2 = E_{b_1} - \frac{q_2}{L} R_{eq} = 11290 \text{ W/m}^2$$

$$\frac{q_{1-2}}{L} = \frac{A_1}{L} F_{1-2} \left(E_{b_1} - J_2 \right) = 2010 \text{ W/m} \quad \Rightarrow q_{1-3} = q_2 - q_{1-2} = 5880 \;\; \text{W/m}$$

$$J_3 = E_{b_1} - \frac{q_{1-3}}{L} \frac{1}{A_1 F_{1-3}} = 17\,370 \text{ W/m}^2$$

Since surface 3 is refractory (adiabatic):

$$J_3 = E_{b_3} \quad \Rightarrow T_3 = 744 \text{ K} \approx 470 \text{ C}$$

12.10 The heat transfer coefficient on the outside of the probe must be found. Assume the depth of immersion of the probe is sufficient so that end effects are insignificant. Assume the film temperature is not much different from the indicated 300C:

$$\nu = 48.85 \times 10^{-6} \text{ m}^2/\text{s} \qquad k = 0.0453 \text{ W/m} \cdot \text{C} \qquad \Pr = 0.684$$

$$\rho = 0.6088 \text{ kg/m}^3 \qquad c_p = 1045 \text{ J/kg} \cdot \text{C}$$

Then the Reynolds number:

$$\text{Re}_D = \frac{VD}{\nu} = \frac{\dot{m}_g}{\rho \pi D_s^2 / 4} \frac{D_t}{\nu} = 334$$

Then from Table 12.8 $C = 0.683$ and $m = 0.466$, and

$$\text{Nu}_D = C \,\text{Re}_D{}^m \Pr^{1/3} = 9.03 \quad \Rightarrow h = 81.8 \text{ W/m}^2 \cdot \text{C}$$

At equilibrium, the energy gained by convection from the gas is equal to the energy radiated to the walls (neglecting conduction along the probe):

$$q_{conv} = q_{rad}$$

$$hA(T_g - T_t) = \varepsilon_t \sigma A(T_t^4 - T_s^4)$$

$$T_g = \frac{\varepsilon_t \sigma}{h}(T_t^4 - T_s^4) + T_t = 363 \text{ C}$$

Multiple Choice Problem Solutions

12.11 **c)**

$$q' = \frac{\Delta T}{L\sum R_{th}}; \quad L\sum R_{th} = \left(\frac{1}{h_i \pi d_i} + \frac{\ln((d_i + 2\Delta r)/d_i)}{2\pi k_{steel}} + \frac{1}{h_o \pi (d_i + 2\Delta r)}\right)$$

$$q' = \frac{120 - 30}{0.0531 + 0.00048 + 0.177} = 391 \text{ W/m}$$

12.12 **d)**

$$q' = \frac{T_o - T_{air}}{1/(h_o \pi d_o)} \Rightarrow T_o = \frac{500}{50 \cdot \pi \cdot 0.036} + 30 = 118 \text{ C}$$

12.13 **b)**

$$q'_{conv} = \frac{T_o - T_{air}}{1/(h_o \pi d_o)} = 283 \text{ W/m};$$

$$q'_{rad} = \varepsilon\sigma\left(T_o^4 - T_{surr}^4\right) = 0.8\left(5.67 \times 10^{-8}\right)\left(353^4 - 298^4\right) = 347 \text{ W/m}$$

12.14 **a)**

$$L\sum R_{th} = 0.2306 + \frac{\ln((0.036 + 0.01)/0.036)}{2\pi \cdot 0.05} = 1.011$$

$$q' = 90/1.011 = 89 \text{ W/m}$$

12.15 **c)**

$$UA = \left(\sum R_{th}\right)^{-1} = 1/0.02306 = 43.4 \text{ W/C}$$

$$NTU = UA/\dot{C}_{\min} = 43.4/84 = 0.517$$

$$\varepsilon = 1 - \exp(-NTU) = 0.403 = q_{act}/q_{\max} = \dot{C}_{water}\Delta T_{water}/\dot{C}_{\min}\Delta T_{\max}$$

$$T_{out} = T_{in} - \varepsilon \Delta T_{\max} = 84 \text{ C}$$

12.16 **c)**

$$A_{1\,fin} = \left[2 \cdot \pi\left(86^2 - 36^2\right)/4 + \pi \cdot 86 \cdot 8\right] \times 10^{-6} = 0.0117 \text{ m}^2$$

$$A_{space} = \pi \cdot 0.036 \cdot 0.008 = 0.000905 \text{ m}^2$$

$$\eta_{total} = 1 - \frac{A_{fin}}{A_{total}}(1 - \eta_{fin}) = 1 - \frac{0.0117}{0.0126}(1 - 0.88) = 0.889$$

$$A_{total}/L = (0.0126/0.016) = 0.788 \text{ m}^2/\text{m}$$

$$LR_{fins} = 1/\left(h \cdot (A_{total}/L)\eta_{total}\right) = 0.0286$$

$$L\sum R_{th} = 0.0531 + +0.00048 + 0.0286 = 0.0822$$

$$q' = \Delta T/\left(L\sum R_{th}\right) = 90/(0.0822) = 1095 \text{ W/m}$$

12.17 **c)**

$$\text{Re} = \frac{4\dot{m}}{\rho\pi D\nu} = 1890 \quad \Rightarrow \text{ Laminar}$$

Nu = 3.66 (assuming constant wall temperture)

$h = 80\ \text{W/m}^2 \cdot \text{C}$

12.18 **b)**

$$\text{Re} = \frac{4\dot{m}}{\rho\pi D\nu} = 7575 \quad \Rightarrow \text{ Turbulent}$$

$$\text{Nu} = 0.023\,\text{Re}^{0.8}\,\text{Pr}^{1/3} = 42.1$$

$h = 918\ \text{W/m}^2 \cdot \text{C}$

12.19 **d)** Use Chilton-Colburn Analogy.

$\text{Re} = 7575;\quad \varepsilon / D = 0.0042 \quad \Rightarrow$ from Moody diagram $f \approx 0.040$

$$\text{Nu} = \frac{f}{8}\text{Re}\,\text{Pr}^{1/3} = 55$$

$h = 1190\ \text{W/m}^2 \cdot \text{C}$

12.20 **b)**

$$\text{Re} = \frac{VD}{\nu} = \frac{2.5 \cdot 0.036}{18 \times 10^{-6}} = 5000$$

$$\text{Nu} = 0.193 \cdot 5000^{0.618} \cdot 0.70^{1/3} = 33$$

$h = 26\ \text{W/m}^2 \cdot \text{C}$

Heating, Ventilation, Air Conditioning and Refrigeration

by Bassem H. Ramadan

Solutions to Practice Problems

Essay Problem Solutions

13.1 Given: $T_d = 85°\text{F}$, $P = 14.7$ psia, and $W = 0.0148$ lbw / lba. From the psychrometric chart:

$T_d = 68°\text{F}$, $W_s = 0.0265$ lbw/lba, $v = 14.1\ \text{ft}^3$ / lba, $h = 36.9$ Btu/lba

$$\therefore \phi = \frac{W}{W_s} = \frac{0.0148}{0.0265} = 0.558 \quad \text{or} \quad 55.8\%$$

13.2 Given: $P = 14.0$ psia, $\phi = 60\%$, and $T_d = 63°\text{F}$. From the steam tables for $T = 63°\text{F}$, we find $P_w = 0.28488$ psia. Hence, $P_{w,s} = P_w / \phi = 0.28488/).6 = 0.4748$ psia. For $P_{w,s} = 0.4748$ we find $T_d = 78°\text{F}$:

$$W = 0.622\frac{P_w}{P - P_w} = \frac{0.285}{14 - 0.285} = 0.021$$

$$v = \frac{R_a T}{P - P_w} = \frac{53.35 \div (78 + 460)}{(14 - 0.285) \times 144} = 14.53\ \text{ft}^3 / \text{lba}$$

13.3 Given: $T_d = 92°\text{F}$, $T_w = 72°\text{F}$, and $P = 13$ psia. From the steam tables we get for $T_1 = 92°\text{F}$, $P_{w,s} = 0.7434$ psia, $h_g = 1101.3$ Btu / lba. For $T_w = 72°\text{F}$ we get $P^*_{w,s} = 0.3356$ psia, $h^*_{fg} = 1052.6$ Btu/lba:

$$W^*_s = 0.622\frac{P^*_{w,s}}{P - P^*_{w,s}} = 0.622\frac{0.38856}{13 - 0.38856} = 0.0308 \text{ lbw / lba}$$

$$W = \frac{W^*_s h^*_{fg} - 0.24(T - T^*)}{h_g - h^*_f} = \frac{.0308 \times 1052.6 - .24(92 - 72)}{1101.3 - 40.07} = 0.026$$

Rearranging Eq.13.1.15,

$$P_w = \frac{PW}{0.622 + W} = \frac{13 \times 0.026}{0.622 + 0.026} = 0.5216 \text{ psia}$$

By interpolation from the steam tables $T_d = 80.9°\text{F}$:

$$\phi = \frac{P_w}{P_{w,s}} = \frac{0.526}{0.7434} = 0.702 \quad \text{or} \quad 70.2\%$$

13.4 Given: $T_d = 85°\text{F}$, $T_w = 68°\text{F}$, $P = 14.7$ psia, and $cfm = 3500$ cfm. Use $\dot{Q}_{sensible} = 0.244\dot{m}_a(T_i - T_s)$ where $\dot{m}_a = cfm \times 60 / v_a$. For $T_d = 85°\text{F}$ and $T_w = 68°\text{F}$, we find $v_a = 13.9\text{ft}^3$ / lba. Hence,

$$\dot{Q}_{sensible} = 0.244\frac{3500 \times 60}{13.9}(85 - 50) = 129{,}000 \text{ Btu / hr}$$

For latent heat gain use

$$\dot{Q}_{latent} = 4880 \times cfm(W_i - W_s)$$

To determine W_s use $T_d = 50°\text{F}$ and $\phi = 60\%$ for comfort reasons; this gives $W_s = 0.0046$ lbw / lba. For $T_d = 85°\text{F}$ and $T_w = 68°\text{F}$ we find $W_i = 0.0108$ lbw/lba. Thus,

$$\dot{Q}_{latent} = 4880 \times 3500(0.0108 - 0.0046) = 105{,}900 \text{ Btu / hr}$$

Finally,

$$\dot{Q}_{total} = 129{,}000 + 105{,}900 = 234{,}900 \text{ Btu / hr}$$

$$\text{Tons of refrigeration} = 234{,}900 / 12{,}000 = 19.6 \text{ tons}$$

13.5 Determine the instantaneous heat gain:

$$\dot{Q} = A[\dot{q}_s \times F_s \times F_t + c_1X + c_2Y]$$

For $T_o = 95°\text{F}$, $T_i = 80°\text{F}$, 40° north latitude, we find at noon,

$$\dot{Q} = 10 \times 6[98 \times 0.87 + 1.0 \times 14 + 0.25 \times 19] = 6240 \text{ Btu / hr}$$

13.6 At 4:00 p.m. the total heat gain is

$$\dot{Q} = 50{,}000 + 500 \times 350 = 225{,}000 \text{ Btu / hr}$$

At 8:00 p.m. the total heat gain is

$$\dot{Q} = 25,000 + 1000 \times 350 = 375,000 \text{ Btu / hr}$$

Hence, the peak load of 375,000 Btu/hr occurs at 8:00 p.m.

13.7 First, find the sensible heat load:

$$\dot{Q}_{sensible} = 25,000 + 1000 \times 245 = 270,000 \text{ Btu / hr}$$
$$= 0.244\dot{m}_a(T_i - T_s)$$
$$\therefore \dot{m}_a = \frac{270,000}{0.244(80-60)} = 55,300 \text{ lba / hr}$$

13.8 Given: $L = 150$ ft, $D = 40$ in., $T_d = 70°$F, $\phi = 50\%$, and $Q = 10,000$ cfm. From Fig.13.2, the friction loss is 0.0355 in. water/100 ft. Hence, the friction loss in the pipe is

$$0.0355\frac{150}{100} = 0.0533 \text{ in. water}$$

Note: in this case, no corrections are needed since the air is at standard conditions.

13.9 The cross-sectional area of the duct at the inlet is

$$A = W \times H = 46 \times 24 / 144 = 7.67 \text{ ft}^2$$

The volume flow rate is

$$Q = 1200 \text{ fpm} \times 7.67 \text{ ft}^2 = 9200 \text{ cfm}$$

The equivalent diameter of the rectangular duct is $D = 35.9$ ". The friction loss is 0.055"water/100 ft:

$$\Delta P_{AB} = 0.055\frac{150}{100} = 0.083 \text{ in. water}$$

a)

$$P_B = P_A - \Delta P_{AB} = 1.0 - 0.083 = 0.917 \text{ in. water}$$

At B, Q = 9200 - 4000 = 5200 cfm

b)

$$V_B = \frac{Q_B}{A} = \frac{5200}{7.67} = 680 \text{ fpm}$$

c)

$$SPR = 0.5\left[\left(\frac{V_i}{4005}\right)^2 - \left(\frac{V_B}{4005}\right)^2\right] = 0.5\left[\left(\frac{1200}{4005}\right)^2 - \left(\frac{680}{4005}\right)^2\right] = 0.03 \text{ in. water}$$

The friction loss in branch BC (for $Q = 5200$ cfm and $D = 35.9$") is 0.018 in. water/100 ft. The total friction loss in branch BC is

$$\Delta P = 0.018\frac{150}{100} = 0.027 \text{ in.water}$$
$$\therefore P_C = P_B - \Delta P + SPR = 0.917 + 0.03 - 0.027 = 0.92 \text{ in.water}$$

13.10 First, calculate the surface area of the walls:

$$A = 2(10\times12+10\times12+10\times10) = 680\ \text{ft}^2$$

The heat gain factor from Table 13.17 is 81 Btu/ft^2/day. For the walls the heat gain is $81\times680 = 55,100$ Btu/day. Next, the internal volume of the room is $10\times10\times12 = 1200\ \text{ft}^3$. From Table 13.20 the number of air changes are 15.4/day. The heat gain from infiltration air is 1.92 Btu/ft^3. Hence,

$$\dot{Q}_{\text{inf}} = 1.92\times15.4\times1200 = 35,500\ \text{Btu / day}$$

The product load is

$$\dot{Q}_{pro} = 1000\ \text{lb}\times0.77\times25\times24/12 = 38,500\ \text{Btu/day}$$

The load due to the people is

$$\dot{Q}_{people} = 750\times24 = 18,000\ \text{Btu / day}$$

The loads are summarized below:

Item	Day	Night
Product	—	38,500
People	18,000	—
Infiltration	35,500	35,500
Transmitted	55,100	55,100
Total	108,600	129,100

The day hourly load is 108,600/24 = 4530 Btu/hr, and the night hourly load is 129,100/24 = 5380 Btu/hr. The night load is higher and should be used when selecting refrigeration equipment.

Multiple Choice Problem Solutions

13.11 **a)** Use $\dot{Q} = UA\Delta T_e$. The solution is summarized as follows:

Surface	Area	U	ΔT_e	$\dot{Q}$
roof	1000	0.54	38	$\dot{Q}_1$
west wall	600	0.34	6	$\dot{Q}_2$
east wall	495	0.34	10	$\dot{Q}_3$
south wall	205	0.34	2	$\dot{Q}_4$

$$\begin{aligned}\dot{Q}_{total} &= \dot{Q}_1+\dot{Q}_2+\dot{Q}_3+\dot{Q}_4\\ &= 1000\times0.54\times38+600\times0.34\times6+495\times0.34\times10+205\times0.34\times2\\ &= 23,600\ \text{Btu/hr}\end{aligned}$$

13.12 **a)** Use $\dot{Q} = A(\dot{q}_s\times F_s\times F_t+c_1X+c_2Y)$. For the east facing window,

$$\begin{aligned}\dot{Q}_{\text{east}} &= 15\times7(127\times0.87+16+0.25\times3)\\ &= 13,360\ \text{Btu/hr}\end{aligned}$$

For the south facing window,

$$\begin{aligned}\dot{Q}_{\text{south}} &= 5\times7(69\times0.87+18+0.25\times16) = 2871\ \text{Btu/hr}\\ \therefore \dot{Q}_{\text{total}} &= 16,200\ \text{Btu/hr}\end{aligned}$$

13.13 **a)** Heat gain per person doing light office work is 400 Btu/hr. Hence, for 20 people the heat gain is

$$\dot{Q} = 20 \times 400 = 8000 \text{ Btu / hr}$$

13.14 **a)** Assume leakage through the door only; plate glass windows have negligible leakage. The crack length is 2(8+4) = 24 ft. Hence, from Table 13.2,

$$\dot{q}_{\text{inf}} = 220 \text{ ft}^3 \text{ / hr / ft}$$

The volume flow rate of infiltration air is 220(24) = 5280 ft³/hr. Hence,

$$\dot{m}_a = \frac{5280 \text{ ft}^3\text{/hr}}{14.35 \text{ ft}^3\text{ /lba}} = 368 \text{ lba/hr}$$

$$\dot{Q}_{sensible} = 0.244\dot{m}_a(T_o - T_i) = 0.244 \times 368(95-72) = 2065 \text{ Btu/hr}$$

$$\dot{Q}_{latent} = 1100\dot{m}_a(W_o - W_i) = 1100 \times 368(.0169-.0101) = 2753 \text{ Btu/hr}$$

$$\therefore \dot{Q}_{total} = 4820 \text{ Btu/hr}$$

13.15 **e)** The mass flow rate is

$$\dot{m}_a = \frac{Q_{sensible}}{0.244(T_i - T_o)} = \frac{50,000}{0.244(72-60)} = 17,077 \text{ lba / hr}$$

To find the volume rate of flow we proceed as follows:

$$\dot{Q}_{latent} = 1100\dot{m}_a(W_i - W_s)$$

$$7000 = 1100 \times 17,077(0.0101 - W_s). \quad \therefore W_s = 0.0097 \text{ lbw / lba}$$

Now, at $T_s = 60°\text{F}$, and $W_s = 0.0097$, we get $v_a = 13.3 \text{ ft}^3\text{/lba}$.

Hence,

$$cfm = \frac{\dot{m}_a v_a}{60} = \frac{17,077 \times 13.3}{60} = 3785$$

13.16 **d)** The sensible load is

$$\dot{Q} = 0.244\dot{m}_a(T_o - T_i)b$$

$$= 0.244\left(\frac{350 \times 60}{14.35}\right)(95-72) \times 0.4 = 3285 \text{ Btu/hr}$$

13.17 **c)** The latent heat load is

$$\dot{Q} = 1100\dot{m}_a(W_o - W_i)b$$

$$= 1100\left(\frac{350 \times 60}{14.35}\right)(0.0169 - 0.0101) \times 0.4 = 4380 \text{ Btu/hr}$$

13.18 **d)** The equipment load is found as follows:

$$\dot{Q}_{sensible} = 2 \times 2200 + 2000 \times 3.413 = 11,230 \text{ Btu/hr}$$

$$\dot{Q}_{latent} = 2 \times 1500 = 3000 \text{ Btu/hr}$$

$$\therefore \dot{Q}_{total} = 14,200 \text{ Btu/hr}$$

13.19 **e)** Given: $T_o = 95°\text{F}$, $T_w = 78°\text{F}$, $T_s = 60°\text{F}$, and $\phi = 60\%$. The load is calculated as follows:

For $T_o = 95°$, $T_w = 78°$, $h_o = 42$ Btu / lba and $W_o = 0.0169$ lbw / lba

For $T_s = 60°$, $\phi = 60\%$, $h_s = 22$ Btu / lba and $W_s = 0.0066$ lbw / lba

$$\therefore \dot{Q} = \dot{Q}_S + \dot{Q}_L = 300[(42-22)+(0.0169-0.0066)28.08]$$
$$= 5913 \text{ Btu / min}$$

$$\text{Tons of refrigeration} = \frac{5913}{200} = 29.6 \text{ tons}$$

13.20 **b)** For this case, $T_s = 65°\text{F}$ and $\phi = 60\%$. $\therefore h_s = 24.1$ Btu / lba and $W_s = 0.0079$ lb. Then

$$\dot{Q} = 300[(42-24.1)+(0.0169-0.0079)\times 33.08]$$
$$= 5460 \text{ Btu/min}$$

$$\therefore \quad \text{Tons of refrigeration} = \frac{5460}{200} = 27.3$$

Power Plants

by Craig W. Somerton

Solutions to Problems

Essay Problem Solutions

14.1

We begin by entering the operating data on the table below:

State	T(°F)	P(psia)
Inlet	240	40
Outlet(ideal)	649	200
Outlet(actual)	680	200

The outlet pressure is given by

$$P_{out} = 5 \times P_{in} = 5 \times 40 = 200 \text{ psia}$$

We use our isentropic relationship to calculate the ideal exit temperature

$$T_{2s} = T_1\left(\frac{P_2}{P_1}\right)^{(k-1)/k} = (240+460)(5)^{(1.4-1)/1.4}$$

$$= 1109°\text{R} = 649°\text{F}$$

The actual work is given by

$$W_{act} = \frac{W_{ideal}}{\eta_s} = \frac{c_p(T_{2s} - T_1)}{\eta_s} = \frac{0.24(649 - 240)}{0.93}$$

$$= 105.5 \text{ BTU/lb}_\text{m}$$

and the actual exit temperature is then calculated as

$$T_2 = T_1 + \frac{W_{act}}{c_p} = 240 + \frac{105.5}{0.24} = 680°\text{F}$$

14.2

We begin by entering the operating data on the table below.

State	*T*(°C)	*P*(MPa)	Fluid Phase	*h*(kJ/kg)	*s*(kJ/kg K)	$\dot{m}$ (kg/s)
Inlet-1	500	15	*Superheat*	*3309*	*6.34*	150
Outlet-1s	*180*	1	*2 Phase* x=0.94	2657	6.34	150
Outlet-1a	*180*	1	*2 Phase* x=0.97	2722	6.46	150
Inlet-2	*180*	1	2 Phase x=0.97	2722	6.46	100
Outlet-2s	*144*	0.4	2 Phase x=0.91	2546	6.46	100
Outlet-2a	*144*	0.4	2 Phase x=0.92	2572	6.49	100

Italicized values from steam tables. Underlined values are calculated.

At the inlet to stage 1, we know the temperature and pressure , so that the remaining properties can be obtained from the steam tables. The ideal outlet conditions for stage 1 are given by

$$s_{O1s} = s_{I1} = 6.34 \text{ kJ / kg} \cdot \text{K}$$

Then from the steam tables we find

$$x_{O1s} = \frac{6.34 - 2.14}{4.45} = 0.94$$

which gives a temperature of 180°C and an enthalpy of 2657 kJ/kg for the ideal outlet to stage 1. For the actual enthalpy for the outlet of stage 1, we use the adiabatic efficiency to yield

$$h_{O1a} = h_{I1} - \eta_s(h_{I1} - h_{O1s}) = 3309 - 0.90(3309 - 2657)$$
$$= 2722 \text{ kJ/kg}$$

From the steam tables we can determine this is a two phase mixture with quality 0.97, entropy of 6.46 kJ/kg·K, and temperature of 180°C. The inlet conditions for stage 2 are the same as for the actual exit of stage 1 and they are entered onto the table appropriately. For the ideal outlet for stage 2 we have

$$s_{O2s} = s_{I2} = 6.46 \text{ kJ / kg} \cdot \text{K}$$

Then from the steam tables we find

$$x_{O2s} = \frac{6.46 - 1.78}{5.12} = 0.91$$

which gives a temperature of 144°C and an enthalpy of 2546 kJ/kg for the ideal outlet to stage 2. For the actual enthalpy for the outlet of stage 2 we use the adiabatic efficiency to yield

$$h_{O2a} = h_{I2} - \eta_s(h_{I2} - h_{O2s}) = 2722 - 0.85(2722 - 2546)$$
$$= 2572 \text{ kJ/kg}$$

From the steam tables we can determine this is a two phase mixture with quality 0.92, entropy of 6.49 kJ/kg·K, and temperature of 144°C. The total power output is then

$$\dot{W}_{tot} = \dot{m}_{I1}(h_{I1} - h_{O1a}) + \dot{m}_{I2}(h_{I2} - h_{O2a})$$
$$= 150(3309 - 2722) + 100(2722 - 2572)$$
$$= 103 \text{ MW}$$

14.3

We begin by entering the operating data on the table below

State	T(°F)	$\dot{m}$ (lb$_m$/hr)
Cold In	220	50,000
Cold Out	1282	50,000
Hot In	1500	50,000
Hot Out	438	50,000

Using the definition of our regenerator effectiveness, we calculate

$$T_{c,out} = T_{c,in} + \varepsilon_{reg}\left(T_{h,in} - T_{c,in}\right) = 220 + 0.83(1500 - 220) = 1282°\text{F}$$

and

$$T_{h,out} = T_{h,in} - \varepsilon_{reg}\left(T_{h,in} - T_{c,in}\right) = 1500 - 0.83(1500 - 220) = 438°\text{F}$$

The internal heat transfer rate is

$$\dot{Q}_{int} = \dot{m}_h c_p\left(T_{h,in} - T_{h,out}\right) = 50000 \times 0.24(1500 - 438) = 12.7 \text{ MBTU / hr}$$

14.4

The energy delivered to the steam is given by

$$\dot{Q}_{stm} = \eta_{furn}\dot{m}_{coal}HV = (0.73)(10)(25\,000) = 182\,500 \text{ kW}$$

The steam exit enthalpy is then calculated

$$h_{out} = h_{in} + \frac{\dot{Q}_{stm}}{\dot{m}_{stm}} = 1594 + \frac{182\,500}{75} = 4027 \text{ kJ / kg}$$

Then from the steam tables we find an exit temperature of 784°C.

14.5

Considering that one mole of fuel reacts, the general reaction equation may be written as:

$$(0.7)CH_4 + (0.3)C_3H_8 + a(TA)O_2 + a(TA)(3.76)N_2 \rightarrow bCO_2 + dH_2O + a(TA - 1)O_2 + a(TA)(3.76)N_2$$

Now proceeding with the element balance,

C: $(0.7) + (0.3)3 = b \quad => b = 1.6$

H: $(0.7)4 + (0.3)8 = 2d \quad => d = 2.6$

O: $2a = 2b + d \quad => a = 2.9$

The chemical reaction equation is then rewritten as

$$(0.7)CH_4 + (0.3)C_3H_8 + 2.9(TA)O_2 + 2.9(TA)(3.76)N_2 \rightarrow 1.6CO_2 + 2.6H_2O + 2.9(TA - 1)O_2 + 2.9(TA)(3.76)N_2$$

To determine the percent excess air, or equivalently the percent theoretical air, we must apply the first law and solve for the percent theoretical air required for the exit temperature. The first law is written

$$H_{products} = H_{reactants}$$

The above equation assumes adiabatic combustion. We also assume that the fuel and air enter at 77°F and write

$$1.6\left\{h_f^o + c_P(T_{prod} - T_o)\right\}_{CO_2} + 2.6\left\{h_f^o + c_P(T_{prod} - T_o)\right\}_{H_2O}$$

$$+ 2.9(EA)c_{P,O_2}(T_{prod} - T_o) + 10.9(1+EA)c_{P,N_2}(T_{prod} - T_o)$$

$$= 0.70h^o_{f,CH_4} + 0.30h^o_{f,C_3H_8}$$

Solving for EA:

$$EA = \left\{0.70h^o_{f,CH_4} + 0.30h^o_{f,C_3H_8} - 1.6\left[h_f^o + c_P(T_{prod} - T_o)\right]_{CO_2}\right.$$

$$\left. - 2.6\left[h_f^o + c_P(T_{prod} - T_o)\right]_{H_2O} - 10.9c_{P,N_2}(T_{prod} - T_o)\right\}$$

$$\div \left\{10.9c_{P,N_2}(T_{prod} - T_o) + 2.9c_{P,O_2}(T_{prod} - T_o)\right\}$$

Substituting in numerical values:

$$EA = \{0.70(-32,210) + 0.30(-44,680) - 1.6[(-169,300) + (8.93)(2723)]$$

$$- 2.6[(-104,040) + (8.01)(2723)] - 10.9(6.94)(2723)\}$$

$$\div \{10.9(6.94)(2723) + 2.9(7.01)(2723)\}$$

which gives a required excess air of 78%. The molar air to fuel ratio is found to be 24.6.

14.6

We begin by writing the chemical reaction equation for the process.

$$C_{65}H_{22}O_{12} + a(1.10)O_2 + a(1.10)(3.76)N_2$$

$$\rightarrow bCO_2 + dH_2O + a(0.10)O_2 + a(1.10)(3.76)N_2$$

The theoretical air has been set to 100% as this will give the maximum product temperature. Now proceeding with the element balance,

C: $65 = b \quad \Rightarrow b = 65$

H: $22 = 2d \quad \Rightarrow d = 11$

O: $12 + 2a = 2b + d \quad \Rightarrow a = 64.5$

Now the chemical reaction equation may be rewritten as:

$$C_{65}H_{22}O_{12} + 71O_2 + 266.8N_2$$

$$\rightarrow 65CO_2 + 11H_2O + 6.5O_2 + 266.8N_2$$

The first law can be solved for the heat transfer to give

$$Q_{released} = H_{reactants} - H_{products}$$

The enthalpy of the products is given by

$$Q_{released} = HHV \cdot MW_{fuel} + 71c_{P,O_2}(T_{reac} - T_o) + 266.8c_{P,N_2}(T_{reac} - T_o)$$
$$- 65c_{P,CO_2}(T_{prod} - T_o) - 11c_{P,H_2O}(T_{prod} - T_o) - 6.5c_{P,O_2}(T_{prod} - T_o)$$
$$- 266.8c_{P,N_2}(T_{prod} - T_o)$$

Substituting numerical values

$$Q_{released} = 6500 \times 994 + 71(7.01)(73) + 266.8(6.94)(73) - 65(8.93)(423)$$
$$- 11(8.01)(423) - 6.5(7.01)(423) - 266.8(6.94)(423)$$

for a heat release of 5.55 MBTU/lbmol of coal.

14.7

a) The thermal efficiency is given by

$$\eta_{th} = \frac{\dot{W}_{net}}{\dot{Q}_{in,total}} = \frac{3}{5} = 0.60$$

b) The maximum efficiency is given by the Carnot cycle efficiency

$$\eta_{\max} = 1 - \frac{T_L}{T_H} = 1 - \frac{25+273}{1500+273} = 0.83$$

c) Applying the first law overall on the system gives

$$\dot{Q}_{intclr} = \dot{Q}_{in,total} + \dot{H}_{gas,in} - \dot{W}_{net} - \dot{H}_{gas,out}$$
$$= \dot{Q}_{in,total} + \dot{m}_{gas}c_P(T_{in} - T_o) - \dot{W}_{net} - \dot{m}_{gas}c_P(T_{in} - T_o)$$
$$= 5000 + (5)(1.00)(25-25) - 3000 - (5)(1.00)(200-25)$$
$$= 1.125 \text{ MW}$$

Multiple Choice Problem Solutions

Questions 14.8–14.15

We begin by performing a cycle analysis which results in the following table:

Node	T(°F)	P(psia)	Phase	h(BTU/lb$_m$)	s(BTU/lb$_m$·°R)	$\dot{m}$ (lb$_m$/hr)
1	800	800	Superheat	*1398*	*1.60*	375,000
2	*550*	300	*Superheat*	*1287*	1.60	48,000
3	*328*	100	*Sat.Vap.*	*1188*	1.60	125,000
4	*193*	10	Sat.Liq.	*161*	*0.28*	125,000
5	*328*	100	Sat.Vap.	*1188*	1.60	202,000
6	*750*	100	*Superheat*	*1404*	*1.82*	202,000
7	*608*	60	*Superheat*	*1336*	1.82	17,340
8	*239*	10	*Superheat*	*1165*	1.82	184,660
9	*193*	10	Sat.Liq.	*161*	*0.28*	184,660
10	193	60	Sub.Liq.	161.15	0.28	184,660
11	*293*	60	Sat.Liq.	262	*0.43*	327,000
12	293	300	Sub.Liq.	262.77	0.43	327,000
13	*417*	300	Sat.Liq.	*394*	*0.59*	375,000
14	417	800	Sub.Liq.	395.75	0.59	375,000

Traverse the cycle.

1-2 **Ideal Turbine** (Stage 1):

$$s_2 = s_1 = 1.60 \text{ BTU / lb}_m \cdot {}^\circ\text{R}$$

The remaining properties can be obtained from the steam tables. The mass flow rate cannot be calculated at this point.

2-3 **Ideal Turbine** (Stage 2):

$$s_3 = s_2 = 1.60 \text{ BTU / lb}_m \cdot {}^\circ\text{R}$$

The remaining properties can be obtained from the steam tables.

3-4 **Heat Load**:

All information known

4-5 **Ideal Turbine** (Stage 3):

$$s_5 = s_3 = 1.60 \text{ BTU / lb}_m \cdot {}^\circ\text{R}$$

The remaining properties can be obtained from the steam tables.

5-6 **Reheater**:

All information known

6-7 **Ideal Turbine** (Stage 1):

$$s_5 = s_3 = 1.60 \text{ BTU / lb}_m \cdot {}^\circ\text{R}$$

The remaining properties can be obtained from the steam tables. The mass flow rate can not be calculated at this point.

7-8 **Ideal Turbine** (Stage 2):

$$s_8 = s_7 = 1.82 \text{ BTU / lb}_m \cdot {}^\circ\text{R}$$

The remaining properties can be obtained from the steam tables.

8-9 **Condenser**:

All information known

9-10 **Ideal Pump**:

$$s_{10} = s_9 = 0.28 \text{ BTU / lb}_m \cdot {}^\circ\text{R}$$

Assuming an incompressible liquid we have

$$T_{10} = T_9 = 193\,{}^\circ\text{F}$$

$$\begin{aligned} h_{10} &= h_9 + v_9(P_{10} - P_9) \\ &= 161 + (0.0166)(60 - 10)\left(144\frac{\text{in}^2}{\text{ft}^2}\right)\left(\frac{1}{778.16}\frac{\text{BTU}}{\text{ft}\cdot\text{lb}_f}\right) \\ &= 161.15 \text{ BTU/lb}_m \end{aligned}$$

11-12 **Ideal Pump**:

$$s_{12} = s_{11} = 0.43 \text{ BTU / lb}_m \cdot {}^\circ\text{R}$$

Assuming an incompressible liquid we have

$$T_{12} = T_{11} = 293\,{}^\circ\text{F}$$

$$h_{12} = h_{11} + v_{11}(P_{12} - P_{11})$$

$$= 262 + (0.0174)(300 - 60)\left(144\frac{\text{in}^2}{\text{ft}^2}\right)\left(\frac{1}{778.16}\frac{\text{BTU}}{\text{ft}\cdot\text{lb}_f}\right)$$

$$= 262.77 \text{ BTU / lb}_m$$

13-14 **Ideal Pump**:

$$s_{14} = s_{13} = 0.59 \text{ BTU / lb}_m\cdot{}^\circ\text{R}$$

Assuming an incompressible liquid we have

$$T_{14} = T_{13} = 417^\circ\text{F}$$

$$h_{14} = h_{13} + v_{13}(P_{14} - P_{13})$$

$$= 394 + (0.0189)(800 - 300)\left(144\frac{\text{in}^2}{\text{ft}^2}\right)\left(\frac{1}{778.16}\frac{\text{BTU}}{\text{ft}\cdot\text{lb}_f}\right)$$

$$= 395.75 \text{ BTU / lb}_m$$

We can now determine the mass flow rates. Starting with the high pressure feedwater heater, we write an energy balance:

$$\dot{m}_{13}h_{13} = \dot{m}_{12}h_{12} + \dot{m}_2 h_2$$

We also note that

$$\dot{m}_{12} = \dot{m}_{13} - \dot{m}_2$$

So we eliminate $\dot{m}_{12}$ from our energy balance and solve for $\dot{m}_2$ as

$$\dot{m}_2 = \frac{\dot{m}_{13}(h_{13} - h_{12})}{h_2 - h_{12}} = \frac{(375{,}000)(394 - 262.77)}{1287 - 262.77} = 48{,}000 \text{ lb}_m/\text{hr}$$

Similarly,

$$\dot{m}_{12} = \dot{m}_{13} - \dot{m}_2 = 375{,}000 - 48{,}000 = 327{,}000 \text{ lb}_m/\text{hr} = \dot{m}_{11}$$

A mass balance on the high pressure turbine gives

$$\dot{m}_5 = \dot{m}_1 - \dot{m}_2 - \dot{m}_3 = 375{,}000 - 48{,}000 - 125{,}000 = 202{,}000 = \dot{m}_6$$

The final mass flow rates are determined form the analysis of the low pressure feedwater heater. We have

$$\dot{m}_7 = \frac{\dot{m}_{11}(h_{11} - h_{10}) - \dot{m}_4(h_{10} - h_4)}{h_7 - h_{10}}$$

$$= \frac{(327{,}000)(262 - 161.15) + (125{,}000)(161.15 - 262)}{1336 - 161.15}$$

$$= 17{,}340 \text{ lb}_m/\text{hr}$$

A mass balance on the low pressure turbine gives

$$\dot{m}_8 = \dot{m}_6 - \dot{m}_7 = 202{,}000 - 17{,}334 = 184{,}660 \text{ lb}_m/\text{hr} = \dot{m}_9 = \dot{m}_{10}$$

With the table filled out, the questions can now be answered.

14.8 **d)** The first law thermal efficiency is defined as

$$\eta_{th} = \frac{\dot{W}_{net}}{\dot{Q}_{in,tot}}$$

$$\begin{aligned}\dot{W}_{net} &= \dot{m}_1 h_1 - \dot{m}_2 h_2 - \dot{m}_3 h_3 - \dot{m}_5 h_5 \\ &\quad + \dot{m}_6 h_6 - \dot{m}_7 h_7 - \dot{m}_8 h_8 \\ &\quad - \dot{m}_{14}(h_{14} - h_{13}) - \dot{m}_{12}(h_{12} - h_{11}) \\ &\quad - \dot{m}_{10}(h_{10} - h_9) \\ &= 118.39 \text{ MBTU/hr}\end{aligned}$$

$$\begin{aligned}\dot{Q}_{in,tot} &= \dot{m}_1(h_1 - h_{14}) + \dot{m}_6(h_6 - h_5) \\ &= 419.48 \text{ MBTU/hr}\end{aligned}$$

which gives a thermal efficiency of 28%.

14.9 **a)** The cycle analysis table shows that the steam leaving the low pressure turbine is superheated vapor so that the quality is not defined.

14.10 **c)** As shown in the efficiency calculation it is 118 MBTU/hr

14.11 **d)** We find

$$\frac{\dot{m}_3 h_3}{\dot{Q}_{in,tot}} = \frac{(125{,}000)(1188)}{419.48 \times 10^6} = 35\%$$

14.12 **d)** We simply sum the mass flows at nodes 2 and 7 to obtain 65,340 lb_m/hr

14.13 **b)** We sum the enthalpy change times the mass flow rate for each pump to obtain 0.94 MBTU/hr.

14.14 **c)** The heat load is heat transfer is

$$\begin{aligned}\dot{Q}_{HL} = \dot{m}_3(h_4 - h_3) &= (125{,}000)(1188 - 161) \\ &= 128 \text{ MBTU/hr}\end{aligned}$$

14.15 **b)** The natural gas flow rate can be calculated from

$$\begin{aligned}\dot{m}_{gas} = \frac{\dot{Q}_{in,tot}}{\eta_{furn} e_{coal}} &= \frac{419.48 \times 10^6}{(0.87)(22{,}000)} \\ &= 22{,}000 \text{ lb}_m/\text{hr}\end{aligned}$$

Fluid Flow

by David C. Wiggert

Solutions to Problems

Essay Problem Solutions

15.1 This is a category 1 problem.

a) The kinematic viscosity of 60° F water is 1.22×10^{-5} ft²/sec. From Table 15.1 we find

$e = 0.00085$ ft. Compute the velocity, Reynolds number, and relative roughness:

$$D = 1.5/12 = 0.125 \text{ ft}, \quad V = \frac{27/(7.48 \times 60)}{\pi \times 0.125^2/4} = 4.9 \text{ ft/sec},$$

$$\text{Re} = \frac{4.9 \times 0.125}{1.22 \times 10^{-5}} = 5.0 \times 10^4, \quad \varepsilon = \frac{0.00085}{0.125} = 0.0068$$

Compute the friction factor:

$$f = 1.325\left[\ln\left(0.27 \times 0.0068 + 5.74 \times (3.2 \times 10^6)^{-0.9}\right)\right]^{-2} = 0.0353$$

(One could also use the Moody Diagram to estimate the friction factor.) Substitute into the Darcy-Weisbach equation:

$$h_L = \frac{0.0353 \times 350 \times 4.9^2}{0.125 \times 2 \times 32.2} = 36.8 \text{ ft}, \quad \text{or} \quad 36.8 \times \frac{62.4}{144} = 15.9 \text{ psi}$$

b) From Table 15.2, assume C = 110. Then substitute known data into the Hazen-Williams formula:

$$h_L = \frac{4.72 \times 350 \times [27/(7.48 \times 60)]^{1.85}}{110^{1.85} \times 0.125^{4.87}} = 38.1 \text{ ft}, \quad \text{or} \quad 38.1 \times \frac{62.4}{144} = 16.5 \text{ psi}$$

Note the difference in the two results; both are acceptable answers for most problems. Solution (a), however, is more accurate.

15.2 This is a category 2 problem. The head loss due to friction is $h_L = 5 \times 144/62.4 = 11.5$ ft. Substitute into the Darcy-Weisbach equation:

$$11.5 = f\frac{600}{4} \times \frac{V^2}{2 \times 32.2}, \quad \text{or} \quad fV^2 = 4.94$$

The Moody diagram is used to estimate f. The kinematic viscosity is ν = 1.22×10^{-5} ft²/sec, and the relative roughness is $\varepsilon = 0.01/4 = 0.0025$. First assume f = 0.03. Then $V = \sqrt{4.94/0.03} = 12.8$ ft/sec, $\text{Re} = 12.8 \times 4/1.22 \times 10^{-5}$ $= 4.2 \times 10^6$, and from the Moody diagram, we find the friction factor to be f = 0.025. Hence the new estimate of the velocity is V = 14.1 ft/sec, and $\text{Re} = 14.1 \times 4/1.22 \times 10^{-5} = 4.6 \times 10^6$. The friction factor remains at 0.025, so that the final velocity is V = 14.1 ft/sec, and the discharge is $Q = 14.1 \times \pi \times 4^2/4 = 177 \text{ ft}^3/\text{sec}$.

As an alternative to the trial and error approach, the empirical formula for a category 2 problem, Table 15.3, can be used since no pump is involved:

$$Q = -0.965 \times \left(\frac{32.2 \times 4^5 \times 11.5}{600}\right)^{0.5} \ln\left[\frac{0.01}{3.7 \times 4} + \left(\frac{3.17 \times (1.22 \times 10^{-5})^2 \times 600}{32.2 \times 4^3 \times 11.5}\right)^{0.5}\right]$$

$$= 177 \text{ ft}^3/\text{sec}$$

The empirical formula gives an answer that is as precise as the trial and error solution.

15.3 This is a category 3 pipe problem. The following data are given: L = 1000 ft, Q = 20/(7.48 × 60) = 0.0446 ft³/sec, and h_L = 210 ft. Substitute into the Darcy-Weisbach equation and simplify:

$$210 = \frac{8 \times f \times 1000 \times 0.0446^2}{32.2 \times \pi^2 \times D^5}, \quad \text{or} \quad D = 0.189 f^{1/5}$$

This relation will be used in conjunction with the Moody diagram to estimate the friction factor. For drawn tubing, $e = 5 \times 10^{-6}$ ft from Table 15.1, and the kinematic viscosity for water at 80° F is $\nu = 0.93 \times 10^{-5}$ ft²/sec. The two parameters required for using the diagram are

$$\varepsilon = \frac{e}{D} = \frac{5 \times 10^{-6}}{D}, \quad \text{and} \quad \text{Re} = \frac{4Q}{\pi D \nu} = \frac{4 \times 0.0446}{\pi \times D \times 0.93 \times 10^{-5}} = \frac{6110}{D}$$

Begin the iterative solution by assuming that $f = 0.015$. Then

$$D = 0.189 \times 0.015^{1/5} = 0.0816 \text{ ft}, \quad \varepsilon = \frac{5\times10^{-6}}{0.0816} = 6\times10^{-5}, \quad Re = \frac{6110}{0.0816} = 75{,}000$$

The updated friction factor read from the Moody diagram is f = 0.019. Repeating the calculations one finds that $D = 0.0855$ ft, $\varepsilon = 6\times10^{-5}$, and Re = 71,000. From the Moody diagram, the friction factor again is f = 0.019, unchanged from the previous estimate. Hence D = 0.0855 × 12 = 1.03 in. One could use either a 1-inch-diameter pipe or a 2-inch-diameter pipe. A 1-inch-diameter pipe will yield a head loss of h_L = 237 ft, and a 2-inch-diameter pipe will yield h_L = 74 ft.

The solution using the formula for a category 3 problem in Table 15.3 is

$$D = 0.66\left[(5\times10^{-6})^{1.25}\left(\frac{1{,}000\times0.0446^2}{32.2\times210}\right)^{4.75} + 0.93\times10^{-5}\times0.0446^{9.4}\left(\frac{1{,}000}{32.2\times210}\right)^{5.2}\right]^{0.04}$$

$$= 0.0868 \text{ ft}, \text{ or } 1.04 \text{ in}$$

Note that the solution is as accurate as using the Moody diagram with the trial and error procedure.

15.4 The kinematic viscosity is $\nu = 1.22\times10^{-5}\ \text{ft}^2/\text{sec}$, the relative roughness is $\varepsilon = 0.00015/0.167 = 0.0009$, and the pipe diameter is $D = 2/12 = 0.167$ ft. Assume a completely turbulent flow regime, so that the friction factor can be approximated by $f = 1.325[\ln(0.27\times0.0009)]^{-2} = 0.019$. Next, write the energy equation from the reservoir surface to the pipe outlet:

$$140 = \frac{V^2}{2g} + \left[\frac{fL}{D} + 2K_{elbow} + K_{entrance}\right]\frac{V^2}{2g}$$

Assume standard elbows so that K_{elbow} = 0.9, and with $K_{entrance}$ = 0.5 from Table 15.4 substitute known data into the energy equation:

$$140 = \left[1 + \frac{0.019\times400}{0.167} + 2\times0.9 + 0.5\right]\frac{V^2}{2\times32.2} \quad . \quad \therefore\ V = 13.6 \text{ ft/sec}$$

We perform one more iteration to determine a better estimate of the velocity:

$$Re = \frac{13.6\times0.167}{1.22\times10^{-5}} = 1.9\times10^5,$$

$$f = 1.325\left[\ln\left(0.27\times0.0009 + 5.74(1.9\times10^5)^{-0.9}\right)\right]^{-2} = 0.021$$

Substituting the new value of f into the energy equation and solving we find that V = 13.0 ft/sec. Hence $Q = 13.0\times\pi\times0.167^2/4 = 0.285\ \text{ft}^3/\text{sec}$. The hydraulic grade line has sudden drops at the entrance and elbows, and a linear drop over the length of the pipe.

15.5 The energy equation for the system is $H_P = (h_L)_{sys} + 150$, or

$$\frac{\dot{W}_P\eta}{\gamma\, Q_{sys}} = (R_1 + R_2 + R_3)Q_{sys}^2 + 150, \text{ in which } R_i = \frac{1}{2gA_i^2}\left(\frac{fL}{D} + \Sigma K\right)_i$$

Compute the R-values:

$$A_1 = \frac{\pi}{4} \times 5^2 = 19.6 \text{ ft}^2, \quad R_1 = \frac{1}{2 \times 32.2 \times 19.6^2}\left(\frac{0.018 \times 600}{5} + 2\right) = 0.00017$$

$$A_2 = \frac{\pi}{4} \times 3^2 = 7.1 \text{ ft}^2, \quad R_2 = \frac{1}{2 \times 32.2 \times 7.1^2}\left(\frac{0.020 \times 1000}{3} + 0\right) = 0.00205$$

$$A_3 = \frac{\pi}{4} \times 4^2 = 12.6 \text{ ft}^2, \quad R_3 = \frac{1}{2 \times 32.2 \times 12.6^2}\left(\frac{0.019 \times 400}{4} + 10\right) = 0.00116$$

Substituting the known data into the energy equation we have

$$\frac{2500 \times 550 \times 0.82}{62.4 Q_{sys}} = (0.00017 + 0.00205 + 0.00116) Q_{sys}^2 + 150$$

which reduces to $18{,}070 = 0.00338 Q_{\text{sys}}^3 + 150 Q_{\text{sys}}$. Solving the equation by trial and error gives

$$Q_{sys} = 99 \text{ ft}^3 / \text{sec}.$$

15.6 Compute the R-values:

$$\text{Pipe 1: } L_e = \frac{2 \times 0.333}{0.02} = 33, \quad R_1 = \frac{8 \times 0.02 \times 233}{32.2 \times \pi^2 \times 0.333^5} = 29$$

$$\text{Pipe 2: } L_e = 0, \quad R_2 = \frac{8 \times 0.02 \times 300}{32.2 \times \pi^2 \times 0.5^5} = 5$$

$$\text{Pipe 3: } L_e = \frac{2 \times 0.167}{0.02} = 33, \quad R_3 = \frac{8 \times 0.02 \times 213}{32.2 \times \pi^2 \times 0.167^5} = 826$$

$$\text{Pipe 4: } L_e = \frac{1 \times 0.333}{0.02} = 17, \quad R_4 = \frac{8 \times 0.02 \times 277}{32.2 \times \pi^2 \times 0.333^5} = 34$$

We next find the system head loss, and then the discharge in each pipe:

$$(h_L)_{sys} = \frac{5^2}{\left(1/\sqrt{29} + 1/\sqrt{5} + 1/\sqrt{826} + 1/\sqrt{34}\right)^2} = 35.5 \text{ ft}$$

$$Q_1 = \sqrt{35.5/29} = 1.11 \text{ ft}^3/\text{sec}, \quad Q_2 = \sqrt{35.5/5} = 2.66 \text{ ft}^3/\text{sec}$$

$$Q_3 = \sqrt{35.5/826} = 0.21 \text{ ft}^3/\text{sec}, \quad Q_4 = \sqrt{35.5/34} = 1.02 \text{ ft}^3/\text{sec}$$

The sum of the four discharges is equal to the system discharge.

15.7 Compute the loss terms for each pipe:

$$\text{Pipe 1: } L_e = \frac{2 \times 0.5}{0.02} = 50, \quad (h_L)_1 = \frac{8 \times 0.02 \times 2550}{32.2 \times \pi^2 \times 0.5^5} Q_1^2 = 41 Q_1^2$$

$$\text{Pipe 2: } L_e = \frac{3 \times 0.333}{0.025} = 40, \quad (h_L)_2 = \frac{8 \times 0.025 \times 1640}{32.2 \times \pi^2 \times 0.333^5} Q_2^2 = 252 Q_2^2$$

$$\text{Pipe 3: } L_e = \frac{7 \times 0.5}{0.018} = 190, \quad (h_L)_3 = \frac{8 \times 0.018 \times 3490}{32.2 \times \pi^2 \times 0.5^5} Q_3^2 = 51 Q_3^2$$

Assume the flow directions shown in Fig. 15.4a, and write the energy relations for each pipe:

$$\text{Pipe 1:}\quad H_B = 65 - 41Q_1^2\,,\quad \text{or}\ Q_1 = \sqrt{(65 - H_B)/41}$$

$$\text{Pipe 2:}\quad H_B = 5 + 252Q_2^2\,,\quad \text{or}\ Q_2 = \sqrt{(H_B - 5)/252}$$

$$\text{Pipe 3:}\quad H_B = 43 + 51Q_3^2\,,\quad \text{or}\ Q_3 = \sqrt{(H_B - 43)/51}$$

The solution is found by iterating values of H_B, computing Q_1, Q_2, and Q_3, and checking the continuity balance at the junction: $\Sigma Q = Q_1 - Q_2 - Q_3$. The results are tabulated below; five iterations are performed until the continuity balance is satisfied within a tolerable limit.

Iteration	H_B	Q_1	Q_2	Q_3	ΣQ
1	43	0.732	0.388	0	+0.344
2	45	0.698	0.398	0.198	+0.102
3	46	0.681	0.403	0.242	+0.036
4	47	0.663	0.408	0.280	-0.025
5	46.6*	0.670	0.406	0.266	-0.002

* Estimated by linear interpolation between the two previous values of H_B and Q.

Hence the discharges are Q_1 = 0.67 ft^3/sec, Q_2 = 0.40 ft^3/sec, and Q_3 = 0.27 ft^3/sec.

15.8 With $D = 8/12 = 0.668$ ft, assuming a constant friction factor, and recognizing that $z_2 - z_1 = 105$ ft, the demand curve can be constructed:

$$H_P = 105 + \left(\frac{0.02 \times 8{,}000}{0.667} + 12.5\right)\frac{Q^2}{2 \times 32.2 \times \left(\pi \times 0.667^2/4\right)^2}$$

$$= 105 + 32Q^2$$

a) Assume Q = 1.0 ft^3/sec, then $H_p = 105 + 32 \times 1^2 = 137$ ft. On the pump characteristic curve of Fig. 15.5, $H_P \cong 175$ ft at Q = 1.0 ft^3/sec. Try another discharge, say Q = 1.5 ft^3/sec; then from the demand curve H_P = 177 ft, and from the pump curve $H_P \cong 170$ ft. Finally, with Q = 1.4 ft^3/sec, from the demand curve H_P = 168 ft, and from the pump curve $H_P \cong 170$ ft. Thus Q = 1.4 ft^3/sec and H_P = 170 ft.

b) From Fig 15.4, the pump efficiency at Q = 1.4 ft^3/sec is approximately 0.7. Hence, the required power input to the pump is $\dot{W}_P = 53 \times 1.4 \times 170/0.7$ $= 18{,}000$ ft - lb/sec, or 32.8 horsepower.

15.9 First compute the head loss in the suction pipe. Neglecting minor losses (no information is given) we have, see Eq. 15.1.1 and Fig. 6.2,

$$h_L = \frac{8 \times 0.02 \times 16 \times (13/7.48)^2}{32.2 \times \pi^2 \times 0.5^5} = 0.78 \text{ ft}$$

a) For water at 80° F, γ = 62.2 lb/ft³, and p_{vap} = 0.507 psi. Substituting into Eq. 15.3.2, the condition for imminent cavitation is

$$NPSH = \frac{(15-0.507)\times 144}{62.2} - 16 - 0.78 = 16.8 \text{ ft}$$

b) With p_{atm} = 12 psi and the same flow conditions for part a), use Eq. 15.3.2 to determine the required z:

$$\Delta z = \frac{p_{atm} - p_{vap}}{\gamma} - h_L - NPSH = \frac{(12-0.507)\times 144}{62.2} - 0.78 - 16.8 = 9.0 \text{ ft}$$

Hence the change in elevation is $16-9=7$ ft.

15.10 The following parameters are given: N_1 = 970 rpm, N_2 = 1,200 rpm, H_1 = 38 ft, Q_1 = 230 ft³/sec, and $D_2 : D_1 = 1 : 2$. Using the similitude relations, Eqs. 15.3.4 and 15.3.5, compute Q_2 and H_2:

$$H_2 = 38 \times \left(\frac{1,200}{970}\right)^2 \times \left(\frac{1}{2}\right)^2 = 14.5 \text{ ft}$$

$$Q_2 = 230 \times \frac{1,200}{970} \times \left(\frac{1}{2}\right)^3 = 35.6 \text{ ft}^3/\text{sec}$$

The power is $\dot{W}_2 = 62.4 \times 35.6 \times 14.5 / 0.85 = 37,900$ ft - lb/sec, or 70 horsepower.

Multiple Choice Problem Solutions

Before answering the specific questions, it is expedient to compute the R-values for each pipe (recall that $h_L = RQ^2$):

$$\text{Pipe 1: } Le = \frac{1\times 0.333}{0.02} = 17 \text{ ft}, \quad R_1 = \frac{8\times 0.02\times 27}{32.2\times\pi^2\times 0.333^5} = 3$$

$$\text{Pipe 2: } Le = \frac{2\times 0.333}{0.02} = 33 \text{ ft}, \quad R_2 = \frac{8\times 0.02\times 533}{32.2\times\pi^2\times 0.333^5} = 66$$

$$\text{Pipe 3: } Le = \frac{2\times 0.333}{0.02} = 33 \text{ ft}, \quad R_3 = \frac{8\times 0.02\times 2,033}{32.2\times\pi^2\times 0.333^5} = 250$$

$$\text{Pipe 4: } Le = \frac{4\times 0.333}{0.02} = 67 \text{ ft}, \quad R_4 = \frac{8\times 0.02\times 1,817}{32.2\times\pi^2\times 0.333^5} = 224$$

15.11 **d)** Compute the discharge and head loss in pipe 2:

$$Q_2 = 5,000\frac{\text{gal}}{\text{hr}} \times \frac{1 \text{ ft}^3}{7.48 \text{ gal}} \times \frac{1 \text{ hr}}{3600 \text{ sec}} = 0.186 \frac{\text{ft}^3}{\text{sec}},$$

$$(h_L)_2 = R_2 Q_2^2 = 66 \times 0.186^2 = 2.28 \text{ ft}$$

15.12 **b)** The flow path is pipe 1 - pipe 2 - pipe 3; there is no flow in pipe 4. The demand curve from A to B is

$$H_P = z_B - z_A + (h_L)_1 + (h_L)_2 + (h_L)_3 = 430 + (3+66+250)Q^2 = 430 + 319Q^2$$

A trial and error solution is employed: assume Q, compute H_P from the demand curve, and compare it with H_P from the pump characteristic curve.

Trial 1: $Q = 10{,}000$ gal/hr $= 0.371$ ft³/sec,

$$H_P \text{ (demand)} = 430 + 319 \times 0.371^2 = 474 \text{ ft},$$
$$H_P \text{ (pump)} = 465 \text{ ft}$$

Trial 2: $Q = 9{,}000$ gal/hr $= 0.334$ ft³/sec,

$$H_P \text{ (demand)} = 430 + 319 \times 0.334^2 = 467 \text{ ft},$$
$$H_P \text{ (pump)} = 466 \text{ ft}$$

15.13 **a)** $Q_2 = 11{,}000$ gal/hr $= 0.409 \text{ ft}^3/\text{sec}$, and from the pump curve, H_P = 460 ft. Write the energy equation from A to C:

$$H_C = H_P - (R_1 + R_2)Q_2^2 = 460 - 69 \times 0.409^2 = 449 \text{ ft}$$

The energy equation from C to D is $H_C = z_D + R_4 Q_4^2$, where H_C is the hydraulic grade line at location C. Therefore the discharge in pipe 4 is

$$Q_4 = \sqrt{\frac{H_C - z_D}{R_4}} = \sqrt{\frac{449 - 445}{224}} = 0.134 \text{ ft}^3/\text{sec}, \text{ or } 3{,}600 \text{ gal/hr}.$$

15.14 **d)** The discharge is $Q = 8{,}000/(7.48 \times 3600) = 0.297 \text{ ft}^3/\text{sec}$. From the pump characteristic curve,

$H_P = 470$ ft, and $\eta = 0.8$. Therefore the required power is

$$\dot{W}_P = \gamma Q H_P/\eta = 62.4 \times 0.297 \times 470/0.8 = 10{,}900 \text{ ft-lb/sec}, \text{ or}$$

19.8 horsepower.

15.15 **c)** First determine the head loss in the suction pipe:

$$Q = \frac{10{,}000}{7.48 \times 3600} = 0.371 \text{ ft}^3/\text{sec}, \quad (h_L)_1 = 3 \times 0.371^2 = 0.41 \text{ ft}$$

Using Eq. 15.3.2, the required pump height is:

$$\Delta z = \frac{p_{atm} - p_{vap}}{\gamma} - (h_L)_1 - NPSH = \frac{(14.7 - 0.256) \times 144}{62.4} - 0.41 - 25 = 7.9 \text{ ft}$$

15.16 **b)** The discharge through the pump is $Q = 12{,}000/(7.48 \times 3600)$ $= 0.446 \text{ ft}^3/\text{sec}$, and from the characteristic curve the pump head is H_P = 454 ft. The energy equation from A to C is $H_P = z_C + p_C/\gamma - z_A + (h_L)_1 + (h_L)_2$. Therefore

$$\frac{p_C}{\gamma} = 454 - 300 - (3 + 66) \times 0.446^2 = 140 \text{ ft}, \quad \text{and } p_C = \frac{62.4 \times 140}{144} = 61 \text{ lb/in}^2$$

15.17 **b)** Write the energy equation from C to D and solve for the pressure at location D:

$p_C/\gamma + z_C = p_D/\gamma + z_D$. (Since there is no flow in pipe 4, the energy equation reduces to the law of hydrostatics.) From Problem 16, the pressure at location C is 61 lb/in^2. Therefore,

$$p_D = 61 + 62.4 \times \frac{300-445}{144} = -1.83 \text{ lb/in}^2$$

15.18 **d)** From the pump curve, for H_P = 460 ft, Q = 11,000 gal/hr, or 0.409 ft^3/sec. Write the energy equation from A to C and evaluate the hydraulic grade line:

$$H_C = H_P - (h_L)_1 - (h_L)_2 = 460 - 69 \times 0.409^2 = 449 \text{ ft}$$

Thus the discharge in pipe 3, between location C and location B is

$$Q_3 = \sqrt{\frac{(H_C - H_B)}{R_3}} = \sqrt{\frac{449-430}{250}} = 0.276 \text{ ft}^3\text{/sec}, \text{ or } 7{,}400 \text{ gal/hr}$$

15.19 **c)** The flow path is pipe 1 - pipe 2 - pipe 4; there is no flow in pipe 3. Write the energy equation from location A to location D:

$$H_P = z_D - z_A + (h_L)_1 + (h_L)_2 + (h_L)_4 = 445 + (3 + 66 + 224)Q^2 = 445 + 293Q^2$$

A trial and error solution is employed: assume Q, compute H_P from the demand curve, and compare it with H_P from the pump characteristic curve.

Trial 1: Q = 10,000 gal/hr = 0.371 ft^3/sec,

$$H_P \text{ (demand)} = 445 + 293 \times 0.371^2 = 485 \text{ ft},$$
$$H_P \text{ (pump)} = 465 \text{ ft}$$

Trial 2: Q = 7,500 gal/hr = 0.278 ft^3/sec,

$$H_P \text{ (demand)} = 445 + 293 \times 0.278^2 = 468 \text{ ft},$$
$$H_P \text{ (pump)} = 475 \text{ ft}$$

Trial 3: Q = 8,000 gal/hr = 0.297 ft^3/sec,

$$H_P \text{ (demand)} = 445 + 293 \times 0.297^2 = 471 \text{ ft},$$
$$H_P \text{ (pump)} = 472 \text{ ft}$$

15.20 **e)** Given data are $Q_1 = Q_2$, N_1 = 900 rev/min. $D_1 : D_2 = 2:3$. Use the similitude relation, Eq. 15.3.4 to compute the new pump speed N_2:

$$N_2 = N_1\left(\frac{Q_2}{Q_1}\right)\left(\frac{D_1}{D_2}\right)^3 = 900 \times 1 \times \left(\frac{3}{2}\right)^3 = 3{,}040 \text{ rev/min}$$

Engineering Economics

by Frank Hatfield

Solutions to Practice Problems

16.1 **d)** Prestige.

16.2 **d)** $1000 \times 1.06^3 = \$1191.$

16.3 **c)** $150/1.08^2 = \$129.$

16.4 **d)** $F = 12{,}000(F/P)_3^{12} - 4000 = \$12{,}860.$

16.5 **c)** $P = 100{,}000(P/A)_\infty^8 = \$1{,}250{,}000.$

16.6 **b)** $500(\mathrm{P/A})_{12}^8 = 500 \times 7.536 = \$3768.$

16.7 **b), d)** $A:\ 300{,}000 + 35{,}000(P/A)_{30}^8 - 50{,}000(P/F)_{30}^8 = \$689{,}000$

$B:\ 689{,}000(A/P)_{30}^8(P/A)_\infty^8 = \$765{,}000.$

16.8 **b)** $\left[1000(P/F)_5^{10} + 2000(P/F)_{10}^{10} + 3500(P/F)_{15}^{10}\right](A/P)_{20}^{10} = \$262.$

16.9 **b)** $10 \times 10^6 = X + 0.1 \times 10^6 (P/A)_\infty^6. \quad \therefore X = 8.33 \times 10^6.$

16.10 **c)** $P = 6000(P/A)_3^{10} - 2000(P/G)_3^{10} = 10{,}300.$

16.11 **e)** By inspection.

16.12 **c)** $P = 200(P/A)^{6}_{10} = \$1472.$

16.13 **b)** $P = 12,000 + 500(P/A)^{10}_{3} + 1000(P/G)^{10}_{3} = \$15,570.$

16.14 **e)** $2P = (F/P)^{i}_{5}P. \quad (F/P)^{i}_{5} = (1+i)^{5} = 2. \quad \therefore i = 15\%.$

16.15 **d)** $54,000(A/P)^{8}_{30} = \$4800.$

16.16 **c)** $10,000(F/P)^{6}_{4} - 3000 = \$9625.$

16.17 **a)** $18,000(A/F)^{4}_{8} = \$1953.$

16.18 **a)** $1000(P/A)^{10}_{\infty} + 1000(P/A)^{10}_{10} = \$16,145.$ Note: for the first 10 years, this accounts for \$2000/yr.

16.19 **a)** $20,000(P/A)^{8}_{6}(P/F)^{8}_{1} = \$85,600.$

16.20 **b)** $e^{0.1} - 1 = 0.10517$ or $10.517\%.$

16.21 **b)** $675 = 50 + 30(P/A)^{i}_{24}. \quad \therefore (P/A)^{i}_{24} = 20.833.$
$\therefore$ by trial and error $i = 0.0116. \quad \therefore 12i = 0.139$ or $13.9\%.$

16.22 **b)** $500,000 + \left[200,000 + 100,000(A/P)^{4}_{3}\right](P/A)^{4}_{\infty} = \$6,400,000.$

16.23 **d)** $400 = 35 + 45(P/A)^{i}_{10}. \quad (P/A)^{i}_{10} = 8.11. \quad i = 4\%. \quad i_n = 12i. \quad \therefore i_n = 48\%.$

16.24 **e)** From preceding solution $i = 4\%.$
$i_e = (1+0.04)^{12} - 1 = 0.601. \quad \therefore i_e = 60.1\%.$

16.25 **d)** $1000 = 300(P/A)^{i}_{4}. \quad (P/A)^{i}_{4} = 3.33. \quad i = 7.7\%. \quad \therefore i_n = 52 \times 7.7 = 400\%.$

16.26 **a)** $A: -16,000(A/P)^{12}_{8} - 2000 + 2000(A/F)^{12}_{8} = -\$5058.$
$B: -30,000(A/P)^{12}_{15} - 1000 + 5000(A/F)^{12}_{15} = -\$5270$

16.27 **d)** $P(A/P)^{10}_{20} = \left[30,000 + 10,000(P/F)^{10}_{4}\right](A/P)^{10}_{8}. \quad \therefore P = \$58,760.$

16.28 **b)** $120,000 + 9000(P/A)^{10}_{6} - 25,000(P/F)^{10}_{6} = \$145,000.$

16.29 **e)** $\left[100,000(A/P)^{10}_{10} + 10,000\right](P/A)^{10}_{\infty} = \$262,700.$

16.30 **d)** $A: 50,000 + 800(P/A)^{8}_{20} = \$57,900$
$B: 30,000 + 500(A/P)^{8}_{20} + \left[30,000 + 400(P/A)^{8}_{10}\right](P/F)^{8}_{10} = \$50,000.$

16.31 **b)** $A: \; -116 + 0.93 \times 206(P/A)_8^{10} = \$906$

$B: \; -60 + 0.89 \times 206(P/A)_8^{10} = \$918$

16.32 **b)** $A: \left[-25{,}000 + 83{,}000(P/F)_3^{10}\right](A/P)_6^{10} - 6000 + 13{,}000(A/F)_6^{10} = \$4260.$

16.33 **d)** $(20{,}000 + P)(A/P)_9^8 - 300 - 5000(A/F)_9^8 = 20{,}000(A/P)_6^8 - 5000(A/F)_6^8.$

$\therefore P = \$7140.$

16.34 **b)** $(23{,}000 - 15{,}000) + (23{,}000 - 32{,}500)(P/F)_N^6 = 0. \quad (1.06)^{-N} = 0.84.$

$\therefore N = 3$ yrs.

16.35 **b)** $-3500(A/P)_5^8 + 12(200 - 50) = \$923.$

16.36 **c)** $5000 + 1.50n = 4.00n. \quad \therefore n = 2000.$

16.37 **b)** $40{,}000x = 500{,}000(A/P)_{15}^8 - 100{,}000(A/F)_{15}^8 + 30{,}000x.$

$\therefore \; x = 5.47.$ Use $x = 6$.

16.38 **d)** $1200 + 40(20 + x) = 90(20) + 50x. \quad \therefore \; x = 20$

16.39 **b)** $65(50) - 65(40) - 1200 = -550$

16.40 **d)** $4500(A/P)_{10}^8 + (18.7 + .1x)75/.9 = 3000(A/P)_{10}^8 + (18.7 + .1x)75/.89$.

$\therefore \; x = 2200$ hr., B

16.41 **d)** $10{,}000 - (10{,}000 - 1300)8/12 = 4200$

16.42 **c)** $D = (4500 - 1500)/3 = 1000$

16.43 **a)** $(300{,}000 - 50{,}000)/7 = 35{,}714$

$300{,}000 - 4(35{,}714) = 157{,}143$

16.44 **c)** $(300{,}000/7)0.5 = 21{,}419$

$300{,}000/7 = 42{,}857$

$300{,}000 - 3.5(42{,}857) = 150{,}000$

16.45 **e)** $300{,}000 \times 2/7 = 85{,}714$

$(300{,}000 - 85{,}714)2/7 = 61{,}224$

$300{,}000 - 85{,}714 - 61{,}224 - 43{,}732 - 31{,}237 = 78{,}092$

16.46 **d)** $(300{,}000 \times 2/7)0.5 = 42{,}857$

$(300{,}000 - 42{,}857)2/7 = 73{,}469$

$300{,}000 - 42{,}857 - 73{,}469 - 52{,}478 - 37{,}484 = 93{,}711$

16.47 **a)** $P = 400S - (35000 + 0.5S^2). \quad S = 1000$ gives $P = -\$135{,}000$

16.48 **b)** $P = 0.\ S = \left(-400 \pm \sqrt{400^2 - 4 \times 0.5 \times 35000}\right) \Big/ (-2 \times 0.5).$

$\therefore S = 400 \pm 300$ units/mo

16.49 **e)** $dP/dS = 400 - 2 \times 0.5S = 0.\ \ S = 400.\ \ P = \$45{,}000$ per month.

16.50 **c)** $U = P/S = 400 - (35000S^{-1} + 0.5S).\ \ dU/dS = 35{,}000S^{-2} - 0.5 = 0.$
$S = 265.\ \ U = \$135/\text{unit}.$

16.51 **c)** $\text{ANEV} = \left[-4000(A/P)_5^{20} - 500\right] - [-1500] = -\$338.$

16.52 **d)** $\text{ANEV} = 0 = \left[-4000(A/P)_5^{20} - 5X\right] - [-15X].\ \ \therefore X = 134$ kg/yr.

16.53 **b)** $D = 4000/3 = \$1333.\ V_2 = 4000 - 2 \times 1333 = \$1333.$

16.54 **b)** $P = 20 \times (0.50 - 0.05) + 15(0.25 - 0.05) - 25 = -\$13.$

16.55 **c)** $P = (20 + 15) \times (0.25 - 0.05) - 25 = -\$18.$

16.56 **d)** $D = \dfrac{(90{,}000 - 15{,}000)}{5} = 15{,}000;\ \ V_2 = 90{,}000 - 2 \times 15{,}000 = \$60{,}000$

16.57 **d)**

$$D_1 = \frac{90{,}000 \times 2}{5} = 36{,}000.\ \ V_1 = 90{,}000 - 36{,}000 = 54{,}000$$

$$D_2 = \frac{54{,}000 \times 2}{5} = 21{,}600.\ \ V_2 = 54{,}000 - 21{,}600 = 32{,}400$$

$$D_3 = \frac{32{,}400 \times 2}{5} = 12{,}960.\ \ V_3 = 32{,}400 - 12{,}960 = 19{,}440$$

$$D_4 = \frac{19{,}440 \times 2}{5} = 7{,}776.\ \ V_4 = 19{,}440 - 7{,}776 = 11{,}664 < 15{,}000$$

16.58 **c)** $\text{ANEV} = -(170{,}000 - 90{,}000)(A/P)_5^{20} - (70{,}000 - 44{,}000)$

$$+ (50{,}000 - 15{,}000)(A/F)_5^{20} = -\$48{,}000$$

16.59 **d)** $\text{PNEV} = (160{,}000 - 100{,}000)(P/A)_5^{20} - (170{,}000 - 90{,}000)$

$$-(70{,}000 - 44{,}000)(P/A)_5^{20} + (50{,}000 - 15{,}000)(P/F)_5^{20} = \$36{,}000.$$

16.60 **a)** $dC/dt = I - \left(L/t^2\right) = 0;\ \ t = \sqrt{L/I}.$

Reader Remarks & Rewards Survey!

Yes, you! ...Your suggestions, comments and errata help us continuously to improve this review. Just fill out, tear, fold, tape and mail this prepaid survey! *Thank you!*

Name: ______________________________

Address: ______________________________

Phone: ____________________ Position: ____________________

"You can tell them I said so!" (check here) ☐

Comments

____

____

____

fold ▶

Any subjects we missed? Where?: ______________________________

Quick Survey

1. Do you think this book is:
(A) Too sketchy
(B) Just about right
(C) Too much material

2. Do you think our problems are:
(A) Too easy
(B) Just about right
(C) Too difficult

3. Do you think our solutions are:
(A) Too sketchy
(B) Just about right
(C) Too much explanation

4. Did you participate in a review course?
(A) Yes Course location/name: ____________________
(B) No

5. Did you use other material in your preparation? If so, please list:
____

6. How long has it been since your undergraduate college graduation?
_____ years _____ recently _____ haven't yet

7. Rank factors in order of influence on your initial appraisal of this book:
_____ Price
_____ Reputation
_____ Written by professors
_____ Presentation of material
_____ Depth/Amount of Material
_____ Other ____________________

Cut or tear here—fold, tape and mail!

Rewards for Errata!

Find mistakes? We happily offer up to $3 for each error found in our *"ME/PE"* (2nd ed.), that has not already been discovered, depending on significance.

error	proposed correction	pg. #	prob. or ex. #

fold ▶

____

Send info to my friend...

I think my friend would like information about your publications. Please send to the following address a catalog and other information relating to the following fields in engineering: Mechanical ☐ Civil ☐ Electrical ☐

Name: ______________________________

Address: ______________________________

______________________________ MESM97

Just fill out, fold & drop this survey card into any mailbox!

NO POSTAGE NECESSARY IF MAILED IN THE UNITED STATES

BUSINESS REPLY MAIL
FIRST-CLASS MAIL PERMIT NO 71 GROVER, MO

POSTAGE WILL BE PAID BY ADDRESSEE

GREAT LAKES PRESS
PO BOX 550
GROVER, MO 63040-9919

fold

Cut or tear here—fold, tape and mail!

fold

Catalog of Titles by Great Lakes Press

Editor and Principal Author: Dr. Merle C. Potter, Ph.D., P.E.

Fundamentals of Engineering Review, new 7th edition, $52.95

The most effective review for the FE/EIT exam. Written by nine PhD teaching professors for the A.M. and General P.M. sessions. Contains all you need to pass —no more, no less! •Includes hardcover reference book which reviews all 12 general subjects; tables and charts; 1100 problems and full solutions; two 8-hour practice exams; NCEES equation summaries. 615 pages.

FE Discipline Review, 2nd edition, $34.95

Reviews the four major discipline P.M. areas: civil, mechanical, electrical and chemical. Covers the most important equations in the *NCEES Reference Handbook*. Includes reviews, practice problems and solutions—plus full 4-hr practice exams for CE, ME & EE. Written by fifteen PhD teaching professors. 362 pages.

Principles & Practice of Mechanical Engineering Review, 2nd ed., $69.95

Full review for the Principles & Practice of Engineering (PE) Mechanical Exam. •The only review written by PhD professors—nine experts in their fields. •Includes reviews of all 15 major subject areas tested; hundreds of excellent exam-simulating practice problems and solved examples. The only concise review. 591 pages. (Practice problems solutions manual, $19.95, 120 pp.)

Principles & Practice of Civil Engineering Review, 2nd edition, $69.95

Full review for the Principles & Practice of Engineering (PE) Civil Exam. •Written by ten PhD professors. •Covers all 16 major subject areas; hundreds of practice problems with solutions. The only concise review. 693 pages. (Practice problems solutions manual, $19.95, 130 pp.)

Principles & Practice of Electrical Engineering Review, $69.95

Full review for the Principles & Practice of Engineering (PE) Electrical Exam. •Written by ten PhD professors. •Concisely covers all major subject areas. 600 pages. (Practice problems solutions manual, $19.95, 110 pp.)

Jump Start the HP-48G/GX, $19.95

Unique user manual written expressly for engineers. Superior to HP manual. Get the most out of your calculator! Makes passing tests much easier! Covers: Using Equations, Solved Problems, Special Applications, Writing Programs, Basic Statistics, File Transfer, and more. 170 pages.

GRE, GMAT & GRE/Engineering Time•Savers™, $17.95

The most authoritative guides to these tests—authors are PhD teaching professors from major universities. •Study same subjects as in other reviews...in *half* the time! •Includes 3 practice tests with solutions and charts to identify your strengths and weaknesses •Complete reviews of all subjects.

QUICK & EASY ORDER FORM

Name (print): ______________________

Street Address: ______________________

____

____

Day ph.: ______________________

Credit Card: VISA MC (circle one)

Number: ______________________

Exp.: __________

Name of Title	Price	Quan.	Multi-Discount!	Subtotal
Fundamentals of Engr	$52.95 x		MULTI-DISCOUNT! Take 15% off each book on orders of 3 or more!	=
FE/Discipline Review	$34.95 x			=
PE Review ME/CE/EE	$69.95 x			=
Solutions ME/CE/EE	$19.95 x			=
JumpStart the HP-48	$19.95 x			=
GRE Time•Saver	$17.95 x			=
GMAT Time•Saver	$17.95 x			=
GRE/Engr Review	$17.95 x			=
Shipping	$4 per item x		N.A.	=
			(MI & MO residents add 6% sales tax.) TOTAL	$

30-day money-back satisfaction guarantee on all GLP titles.
Mail this order form, credit info, & check or money order to:

Great Lakes Press POB 550 Grover, MO 63040

Call 1-800-837-0201 / fax 1-314-273-6086
www.glpbooks.com / custserv@glpbooks.com

MP/SM-97